PNEUMATICS IN INDUSTRY

SYSTEM DESIGN AND VIBRATION ISOLATION

E. B. PATERSON

M.C., F.I.M.C., F.I.M.H.
Consultant in Applied Pneumatics

McGRAW-HILL BOOK COMPANY Auckland

New York St Louis San Francisco Auckland Bogotá Hamburg
Johannesburg Lisbon London Madrid Mexico City Montreal
New Delhi Panama Paris San Juan São Paulo Singapore
Sydney Tokyo Toronto

ISBN 0 07 008309 6

Printed in Hong Kong by Everbest Printing Co. Limited.

Sponsoring Editor: Stuart Laurence
Copy Editor: Daphne Rawling
Designer: George Sirett
Technical Illustrator: E.B. Paterson

Contents

v

Preface

This book attempts to present the refinements of pneumatic control, and to highlight those areas in the use of pneumatic power and control equipment in which productivity, quality control and profitability are directly affected for good or ill in any industry. These areas include control circuit design, selection of equipment, staff training, workshop equipment and, last but not least, effective isolation of vibration which, when uninhibited, is so damaging to the efficiency and well-being of plant, equipment, buildings and personnel associated with industry.

To appreciate fully the implications of much of the contents here it is necessary to have some familiarity with the contents of *Practical Pneumatics: An Introduction to Low Cost Automation*. Indeed this book has been written for those who wish to take pneumatics beyond the limits of the distance covered by *Practical Pneumatics*. As was the case with that book, this volume is an attempt to present the practical engineer with simple, usable tools with which he can deal with the day-to-day problems as they occur. By following closely the guidelines described herein, the end result should give cause for satisfaction. That this should be, is the earnest aim of the author.

A book of this nature could never be looked upon as the product of one person or even a group of people. It represents the ideas and knowledge of many who, over the years, have tested them in practical fashion. Before recognising such ideas as valid guidelines, it has been necessary to relate back to basic principles and the common laws of nature. Again, this has not been the work of any one individual.

Thus, to attempt to thank all those who may be regarded as contributors is an impossible task. I am grateful to all with whom it has been my good fortune to experience some association which has contributed some part of the whole. I trust they will accept this as evidence of my indebtedness.

There are, however, some to whom especial thanks should be recorded. Two companies have been particularly generous in allowing the liberal use of material and illustrations from their own publications and knowledge from their own research and technical findings. First, Martonair International Ltd, together with the late George Godwin, who opened many doors for me, and his two colleagues, Russell Jennings and Fred Barnes, who, throughout several decades, have been so helpful in all matters pertaining to pneumatic power and control.

Second, in the field of vibration isolation, the Firestone Industrial Products Company has been equally generous in allowing the use of knowledge, materials, and illustrations which have been, in the main, confined to their own publications. I am especially grateful, in this connection, to Garry Reynolds and Bill Grepp. Both of them have been generous in the extreme. Bill Grepp has not only shared his thirty years or more of pioneer work, which has established him as a world authority, but also was kind enough to check Chapter 11 in this book. Without these two companies and those named here this book could not have been written.

In addition, I am grateful to several others whose cooperation and encouragement has been so much appreciated over the years —Arthur Salek, Neil Ellis and his company, MacEwan's Machinery Ltd, Athol Drinnan, Ted O'Hara, Roger Watson, Wayne Larsen, Ken Hookway, David Catt, John Chandler, Robert Vercoe and Jock McGregor, to name a few. More recently I have had cause to be grateful to Dr A.C. Tsoi and Jim Sharp, whose constructive criticism was of help in the final shaping of the book.

Finally, the patience, literary ability and encouragement of my wife, Pat, has sustained me in a task which, at times, I could cheerfully have abandoned (and she, me!). I am grateful to her and to all.

E.B. PATERSON
Waiheke Island, NZ

Introduction

The cornerstone of good engineering is simplicity. In production engineering the retention of simplicity is as essential now as it has ever been. The approach to engineering problems associated with the manufacture and design of the plant required, together with the practical answers to the problems as they take shape as manufacturing units on the factory floor, need to be simple.

The growth in technology has brought us into the age of specialisation so quickly that it has become increasingly difficult to maintain a balanced view of any problem when surrounded by so many who know so much about so little.

The specialist in many cases has so much to master in his own field that he has little time to become acquainted with even the barest details of any alternatives to procedures with which he is familiar. The pros and cons of any alternatives are rarely examined when there are so many practical problems to be mastered in his chosen medium and methods.

Production engineers would often rather turn their attention to, and are better equipped to apply, computer drive to a $50 000 000 factory than control effectively the speed and thrust of a single, double-acting, cushioned air cylinder. There is also some intellectual snobbery which sees little personal prestige in associating with simple things.

As a result of this trend towards specialisation, enthusiasts with a great deal of knowledge in a narrow field will often apply unnecessarily sophisticated systems to control relatively simple processes, when a simpler and less costly means could have controlled the process equally as well. Not only could the simpler system have carried out the work satisfactorily but also, because of its simplicity, its maintenance would present no problems to less specialised engineers whose background and general knowledge would be wider than those of their specialist counterparts.

As an example, take the case of a microprocessor applied to control a manufacturing unit in which five air cylinders perform a cycle of operation with each of the five movements following one after the other in a fixed sequence. The microprocessor would be ensuring correct sequential control, with possibly a short fail-safe cycle of operation to be carried out in the event of an emergency.

To do this, all feedback signals to the processor would be electrical, as would all signals from the processor. Since all the power units would be pneumatic, controlled by pneumatic five-port relay valves, all signals calling for movement would have to be converted from electrical to pneumatic by solenoid pilot valves. Any fail-safe provision would have to take into account the possibility of an electrical power failure. This would entail the probable use of an emergency store of compressed air in a reservoir so that fail-safe action would be performed pneumatically. Sensing devices would be likely to be electrical microswitches or electrical proximity switches.

Controlling the same unit pneumatically, the cylinders would remain unchanged. The sensing devices would be the pneumatic equivalents such as microvalves, pneumatic proximity switches or simple, low-pressure air-jets occluded at appropriate positions to signal the arrival or correct positioning of some part on the machine of the product being processed. Since these sensing devices give out pneumatic signals, there would be no need for the solenoid pilot valves. If the disposal of signals no longer needed was considered when arranging the sequence, small five-port valves could be used and the cascade system of sequential control applied.

In this instance, the microprocessor would clearly be wasted. Only a very small portion of its total potential would be put to work. Further, its use would unnecessarily complicate the control system by adding both itself as a complex unit and a number of unnecessary solenoid pilot valves. This, in turn, would add considerably to the cost, both initially and thereafter, through the very different type of maintenance required.

This example may seem a little exaggerated, but it is typical of several observed in various manufacturing industries. Usually such cases are originated by an engineer who has had specialist training. Knowing no alternatives and determined to make sure his company's plant is up to date, he could, in his enthusiasm, persuade non-technical directors that the shareholders were benefiting from the very latest in advanced technology. In fact, with the best of intentions he has added capital costs which will be difficult to retrieve from the output of that particular unit.

The microprocessor represents a decided step forward in its particular field and there are many applications for which it is ideally suited. The point to be emphasised, however, is that every process, every cycle of operation, has a particular medium and control system which would carry out the work more cheaply, more simply and more effectively than any other alternative. It is the obligation of the engineer to explore in a knowledgeable fashion the various alternatives before making the decision as to which will be used.

In many modern processes there is much which ideally can be done by units in which a number of movements follow one after another in a sequence of operation which may remain unchanged over years. For many such units the ideal medium, both for power and the control system, is compressed air.

Again, in many manufacturing plants there are operations of one or two movements only. Often these movements are performed by a person rather than by a machine. In such cases both output and quality control are directly dependent on the person's activities. Even with a highly skilled operator, output and quality control can vary enormously throughout the day's work, as the operator tires, is refreshed by a break, then tires again. In such areas, the air cylinder will provide a tireless muscle and a consistency of performance which will increase both output and quality control.

Productivity is a term which conveys many different meanings to different people. However, if we consider it as expressing the antithesis of waste (whether it be waste of energy, waste of time, waste of space, or waste of materials and resources generally) it will serve our purpose here well enough in examining the practical use of pneumatic equipment.

Quality control is a necessary adjunct in any production plan. It might be described as the ability to maintain consistency at a pre-set level in regard to both the material and the process by which the raw material is transformed into the finished product.

The practical applications of pneumatics has much to offer towards these objectives. *Practical Pneumatics: An Introduction to Low Cost Automation* dealt with the basics of practical applied pneumatics. This book is an attempt to refine and apply those basics with the objectives of simplicity, productivity and quality control—all equally important if the product is to be produced economically and find and maintain a worthwhile place in the market.

Chapter 1

Circuit design for a cycle of operations

A cycle of operations may be a single movement taking place every time a predetermined set of conditions occurs; or it may be a predetermined chain of events which is repeated in various patterns to meet the requirements of the conditions related to the process controlled by these events. These events themselves could be anything from the turning on of a water valve to the transfer of material from one point to another, the movement of a drill or the closing of a hopper gate.

To set up the means of producing a train of events which will react on the material to produce the desired result, we need components which may be grouped under three general functional headings:

1. **Prime movers** which will directly apply the energy or force required for each individual event.
2. **Sensors** to sense the change of conditions taking place in the process and provide "feedback". This term implies registration of the change of condition and provision of a signal back to the logic system signifying the change which has taken place so that the next event in the process may take place.
3. Some sort of **logic system** which will sort out the messages from the sensors and apply signals for further action at appropriate times.

Technology now offers an almost bewildering array of components under all three groups, so much so that it is difficult to rationalise the picture and make an intelligent selection of the overall control system and its components which will ideally suit the process and its environment. Yet this is the task faced by every engineer who accepts the responsibility for the design of a new project, large or small.

Obviously, selection of the most appropriate system demands at least an acquaintance with the general principles and methods employed in the possible alternative systems. Also, the engineer making the selection must have some sort of criteria against which to match the systems examined in a search for the most suitable. Each of the alternatives will have a particular set of conditions with which it alone is the nearest to the ideal for the purpose. Thus there must be a very clear idea of the function required and the overall conditions in which the selected system will have to work. Identifying all these last details is not only one of the most important but also often one of the most difficult aspects of the task.

In chronological order the action leading up to the detailed design of a suitable control circuit for any project entails the following.

1. The preparation of the criteria required for the project in respect to:
   (a) the functional aspects of the project,
   (b) the environmental aspects of the project.
2. Selection of the most suitable medium or combination of media for both power and control when comparing the two criteria established in 1.
3. Selection of the most suitable method and system together with related components which are indicated as the logical outcome of the factors established in 1 and 2.

At no stage in the planning of the project is there room for any preconceived ideas as to how the job will be done. Experience may shorten the time taken to establish all the relevant factors, through knowing where to look for some of the less obvious related aspects and in better developed ability to recognise some of the signs. Unfortunately, experience is not something acquired from a textbook or overnight on the job.

Often, at intervals throughout the exercise, it is worthwhile to bring together several engineers whose individual specialised knowledge covers possible alternatives. These alternatives may then be considered and compared. The pros and cons can be discussed and written down.

It will often be found that the boundaries within which any particular medium, system or method best applies become fairly clear. However, even the most experienced engineer cannot afford to take for granted that these boundaries are inflexible or even remain the

same for any length of time. Constant development is a feature of all techniques. What is right today may be only second best tomorrow. This applies both when comparing the methods employed by the various alternative media and when comparing the various methods used within an individual medium.

The three stages of preparatory work involved in a new project are all aimed at achieving the maximum productivity from the minimum initial and continuing cost of equipment, when related to the value of the product resulting from the new plant. In this context the terms "cost" and "value" are by no means confined to their use as monetary terms. To confine your thinking to such a short-term view usually spells disaster in the longer term. To take one example, a high capital value/low-level labour plan is not always the most economic in the long run—the reverse is often the case, as many case histories show. There can be no preconceived ideas of value—each project must stand on its own to be evaluated and assume tangible form as related to its own particular overall conditions.

Major pneumatic factors affecting productivity

Productivity is affected directly by many diverse conditions. In the field of pneumatics it starts in the design of the control circuits and follows on through almost every part of the process of transforming an idea into a working machine. Included in this is the arranging of the sequence of movement, the selection of working control and power components, the degree of control achieved with each individual movement, the layout of working equipment, the ease of identification and accessibility of the working components for servicing, together with the availability of spare parts for the components selected. And all of these factors make up only one side of the picture.

The other side is the environment in which the equipment will have to work. Included in this are the quality of the air supply, the compatibility of the total concept of the system with the workforce who will be associated with it, the amount of training the maintenance staff have in pneumatics, the facilities at their disposal in the workshop to enable them to carry out maintenance, whether preventive or emergency, with the minimum of down time and associated loss of production—all of these factors and many others combine to enhance or restrict the productivity of the project when it takes its place on the factory floor.

Matters determining the line of approach

The decision to build a new machine or modify an existing part of the plant will be the result of research into the market potential of the product. This research should have established rough boundaries within which the designer can work in relation to both cost and rate of production.

Functional aspects of the project

The first consideration in design planning is to establish precisely what has to be done to the raw material or workpiece to change it into the finished product. Every known detail should be written down and studied to ensure that nothing is omitted from the process. All the action required should be noted carefully so that there is no doubt about the precise sequence of events. The type of action required should be studied carefully. If it is a movement, the feasible range of operating speeds and thrusts need to be studied. Where a number of events follow in succession, the time which can be allowed between each should be noted. These considerations will be important when a decision is made as to the types of signals and the medium suitable for such. If, as is often the case in an entirely new development project, there is an element of the unknown in any particular section of the operation, that section

should be noted so that room in the planning may be left for the insertion of alternatives and possible changes in the concept at that point.

From this detailed study of the functional aspects of the project, a working specification of a relatively flexible nature can be drawn up.

Environmental aspects of the project

In any major project, the environmental impact of the project has become an important determinant of the design. It is equally important in small projects. The long-term economics of any project derives incalculable benefit from time spent in studying the environmental aspects before attempting the final design.

The best method to employ is to pose questions—and write them down. Then set about finding the answers, and write them down too. In this part of the exercise, there will be two alternatives to take into account, each with its own set of questions and answers:

1. Will the process be fully automated?

OR

2. Should there be a degree of human labour interspersed through a series of semi-automatic steps?

Some of the questions appropriate to the first question would be:

1. If fully automatic, what sort of interest will it gain from those who will look after it?
2. If there is a doubt in respect to interest, would it be better to build in an element of human responsibility to maintain interest in the overall efficient working of the machine?
3. What sort of quality control is required?
4. Can the degree of consistent repetitive precision be achieved within the cost boundaries laid down?
5. Do we need checkpoints for human control of quality in stages through the process, or is it safe to rely on automatic checking and, where necessary, rejection?
6. Is the working environment compatible with the type of equipment we shall have to use?
7. Do we know what the working environment

might be in respect to dirt, fumes, noise, temperature extremes, excessive moisture, etc., against which the process may need protection?

8. Do we know the standard of skills and training normally to be found in the operators and maintenance staff who will be associated with the machine in its working life?

9. Are any special safety factors required?

10. What factors must be considered in relation to emergency stops?

Some of the questions appropriate to the second question would be:

1. Are there good reasons for creating a high labour content apart from the actual efficiency of the machine itself?

2. Is there a labour force available?

3. Will we get better quality control by having some of the operations dependent on human judgment and skills?

4. Will it generate a more stable workforce in the factory by creating a degree of interest and job satisfaction through dependence on human skills?

5. If reliance is to be placed on human skills, what sort of instruction will be required by the staff who operate the machine? Will such instruction fall within the capabilities of those likely to be available? Will the instruction required be compatible with their educational and cultural background?

6. If the process is to be divided into a series of steps, each step carried out by a semi-automatic unit employing an operator, can the output of each unit be conveniently matched with those of the other units in the group so that a steady balanced flow can be established throughout the process?

There will probably be many other questions which should be asked but these will serve as an indication.

Having established the general criteria, the question arises of which of the media should be used for first the power units then the control system. The possible alternatives should be examined and the pros and cons of each noted so that the final decision will be based on the widest possible knowledge. Here, as suggested earlier, it is often worthwhile to draw into the discussion several engineers, each with a different specialised background. If the power units are predominantly pneumatic, keeping the control system in the same medium would tend to appear the more attractive alternative, provided that the types of sensors required fall within the capabilities of the pneumatic range of sensors. This would avoid the use of solenoid pilot valves to transpose electrical signals to pneumatic signals every time a power unit was required to perform. On the other hand, other considerations may outweigh what at first sight seems the best way of handling it.

For instance, a steel mill rolling steel rod may use air cylinders for most of the action. However, when the smallest diameter rod reaches its near final stages of rolling, it can be travelling through the rollers at about 150 km per hour. When the required length for the roll is reaching the point where it will pass through the cutters, the length of time taken for a pneumatic signal to initiate the cutting cylinders, with the steel passing through at over forty metres per second, could give enough time for variables in roller speeds, rod slip, etc., to contribute to relatively inaccurate cutting and variable lengths of rod in the roll. This is just one aspect of a complex process, in which furnace temperature and the factors contributing to that, plus roller speeds, etc. (much of which relies on electronic sensing devices) are all tied together in what is termed a closed loop system. In this case, the obvious choice for the main control system would not be pneumatic as most of the signals are electronic, much of the control system relies on electronic proportional devices and many of the sensors are electronic with no pneumatic equivalents.

On the other hand, the case of a cement distribution plant, with large silos containing a variety of cements and additives, offers an interesting contrast to the steel mill. In this case, the distribution fleet of cement transporters was filled at filling points where each load was selected by the operator from various combinations of silos. The combination depended on the individual order about to be delivered by the transporter. All the hopper gates and conveyors, or air-slides, were pneumatically operated. The control panels and logic system, however, were electrical. From the central control board, electrical signals were sent to solenoid-operated pneumatic relay valves which operated the air cylinders. Sensors detecting whether the hopper gates were open or closed were electrical. Because of the fine cement dust which hung in the air throughout the plant, permeating almost everything, electrical contacts had no chance of remaining clean and effective. Thus, at least every two or three days a major breakdown would occur. A hopper gate would fail to close and the transporter would be buried in many tonnes of cement. A conveyor filling a silo would fail to stop and the silo would overfill, spilling large quantities of cement on the surrounding area. The owners of the plant were reluctant to change the control system to an all-pneumatic system because of a mistaken conception of the length of time taken for a pneumatic signal to reach its destination. In the plant, signals had to travel upwards of seventy to eighty metres from the control cabinets to some of the hopper gates. The owners did not realise that signals would often take no more than 0.8 second not only to travel to the valve concerned but also to change over the valve receiving the signal. They eventually realised that, provided they selected pneumatic equipment with the necessary consistency of response, such minor time lags could be anticipated and compensated. They also realised that the pneumatic equipment, with the simple means of protection normally used in any plant, would operate reliably under the particularly unpleasant conditions. Finally, the changeover in the control system was made. A check was maintained on performance for the next two years during which not a single malfunction of the system took place.

In both the cases quoted, final satisfactory performance was achieved when the function of the plant, the environmental aspects of the plant and a working knowledge of the various alternative control systems possible were brought together and matched with each other. Certainly, in the case of the cement depot it took a little longer to find out about the alternatives, but when this was done, the result was a reliable, efficient plant.

Assuming that all the relevant facts finally point to the use of compressed air for both power and control, there are still several alternatives to be investigated before a final decision is made and the actual drawing up of a circuit diagram started.

As *Practical Pneumatics* pointed out, in a pneumatic sequence of operation the designer often has the problem of disposing of a signal no longer required before another signal can be applied to the opposite pressure end of a double pressure-operated valve. A simple and effective method of dealing with this, and also a means of identifying the type of sequence in which this occurs, was described in that book. However, there are a number of other methods of dealing with this problem, all of which have some merit under certain circumstances. Therefore, it is always wise to consider which of these methods is most appropriate to the particular problem in hand.

The criteria covering the selection of the method to be used will include:

1. simplicity
2. economy of equipment
3. ease of fault finding
4. proven reliability
5. proven life expectancy
6. availability of components and spares
7. attributes which may be especially suited to the particular problem under review
8. training required by maintenance staff

Because of the problem of opposing signals, it is worthwhile establishing first the circuit required for the underlying sequence of events. This basic structure of the circuit might be regarded as the skeleton on which the flesh will be built. The "flesh" is discussed over the next few pages.

Details are needed of the requirements of each individual movement in respect to load, speed, stroke, cushioning, together with any features peculiar to that particular movement. For example, to conserve air it should be known if a large air cylinder will operate on normal lines pressure on its load handling stroke and lower pressure on its return, unladen stroke. If so, the required pressure reducer would be drawn, and the relay valve for the cylinder drawn in the manner required to show the respective flow directions of high and low pressure air. Speed would be indicated at that stage by the sizing of the relay and an indication of this shown on the circuit. Size of the pipe between relay and cylinder would be shown on the drawing, together with the flow regulators, drawn to show in which direction they control the speed.

It is of the utmost importance that each individual movement should receive careful consideration. The success or failure of any project, large or small, relies on each movement working. There are cases of very large industries working at reduced output for months or even stopped altogether because of the malfunctioning of one single air cylinder. In a large steel mill, the new corrugated iron section operated at half design speed for six months after it had been commissioned. Many thousands of dollars of forward orders were lost, and supply contracts with penalty clauses caused heavy losses in penalty payments. No one had taken the trouble to work out the cushioning required for one large cylinder responsible for part of the guillotine operation. The guillotine was unable to cut the corrugated iron sheets at any speed above half the design speed. The total output was therefore reduced to half until, six months later, the real cause of the trouble had been correctly diagnosed and rectified.

In another example, the construction of the main dam for a large hydroelectric scheme was completely held up for two days. This was caused by the malfunctioning of an air cylinder with a 32 mm bore and 100 mm stroke. The very large concrete batching plant relied on this small cylinder to inject one of the essential additives to the concrete mix. The concrete supply contractors were bound to supply concrete which was certified by approved methods as having all the essential ingredients in the mix. No one had given this small movement the thought and care it required in the design stage of the concrete batching plant—a new one for the project. Thus when the batching plant started up for its first day's work—weeks overdue and then on penalty clause payments—the additive cylinder could not perform

its intended function. There was no production of concrete for a further two days. The large labour force involved was idle for that time; another two days were added to the delay before the hydroelectric scheme was finally producing its quota of power.

One of the dangers of any project which appears to be complex in any way is that those involved may become so excited over the very complexity of it all that they lose sight of the fact that nothing is really complex. Everything can always be broken down into small simple items and each of those is as important as the whole. The chain is only as strong as its weakest link—axiomatic but often overlooked in the stress of the moment.

Time delays required throughout the cycle of operation will require some study as to their purpose, how many times they may require changing within a given period, degree of accuracy required, intervals between delays as compared with actual length of delays, etc. This can be obtained by referring back to the notes of the initial investigation into the complete operation. From this information a decision must be made: whether it would be better to use a factory-made timing device with a calibrated dial or to design the appropriate simple circuit using standard components. If the simpler and usually less expensive circuit with ordinary components were used, a decision would be required as to which of the methods described in *Practical Pneumatics*, pp.92–4, would be appropriate. Using such methods, accuracy of a high order is practicable. Usually maintenance costs throughout the life of the machine will be found to be lower using these latter methods than with the factory built timer. As a rule, also, spares for the components used will be readily available as they will be needed throughout the system for other purposes, whereas spares for the factory built timer may not be so readily available.

Fail-safe and emergency stop requirements are an important part of the "flesh" which must be added to the circuit at this stage. These require a good deal of thought before a practicable answer is forthcoming which can provide essential safety measures without submerging the whole circuit under a superstructure of fail-safe subcircuits. A great deal of time can be wasted in speculation on the "what-ifs", apart from the danger of including so many sub-circuits to react to possible changes of condition that the original purpose of the machine can fade into obscurity. Many of the "what-ifs" can spring from lack of confidence in the reliability of components themselves. If all reasonable precautions and care are taken in respect to providing normal working conditions and in selecting equipment with some proven record, there should be no need to provide excessive back-up units. The question of fail-safe and emergency action is dealt with in detail in a later chapter of this volume.

The circuit will now have reached a stage where a check can be made against the cycling time of the operation set down as the maximum in the original specification. This check provides an opportunity of rechecking the sequence of events to ensure that each event is a productive one. The unproductive movements, such as return of unladen cylinders preparatory to the next working strokes, resetting of valves, etc., should all occur during some productive event. To maintain quality control, a check should be made over the circuit to ensure that there will always be a consistency of performance throughout the system—pipes large enough to avoid "air starvation" of any power unit or section of the control system at any given part of the cycle. For example, undersized pipe may produce an unreliable signal which has to take place while a large air cylinder is moving. In moving, the cylinder could be producing pressure drops below the safe working pressure for the reliable signal. Components which are marked down for selection should now be checked for their repetitive consistency in response to a signal. This is usually a matter of seal and general valve design, as pointed out in *Practical Pneumatics*.

The question of lubrication, length of pipelines from control cabinets to equipment mounted on the machinery, and practical matters to make the installation simple and effective will require attention and thought. Instructions concerning the installation and operation of the process or machine will need to be prepared before the circuit is placed in the hands of those who will build, install and commission. The layout, accessibility of all components and general appearance should also be considered so that when the tangible evidence of all the preparatory work appears as a working unit it can look forward to a long and useful life.

Throughout the setting up of the whole project, from its initial conception until it is running in production on the factory floor, it must always be borne in mind that every decision made will have either a favourable or an adverse influence on the unit's final productivity and its inbuilt means of controlling the quality of the finished products it creates.

It is not sufficient, however, merely to place a well-conceived, well-executed production unit on the factory floor and expect it to attain any measure of productivity or quality control. From the time the unit embarks on its productive life, the following factors will have a direct bearing on how well it will perform its designed function:

1. The degree of understanding of both its capabilities and limitations which operators and engineering maintenance staff have *before* it goes into production, coupled with general understanding of its normal functioning or what makes it work.
2. The attention given to its related working environment. For example, before the machine is installed, the quality of the factory compressed air supply must be checked to make sure that it is clean, free from moisture and compressor oil, and of sufficient volume and consistent pressure to ensure correct operation of all pneumatic components. At design stage, the presence of any unusual conditions in the proposed working area must be established. Corrosive elements, excessive heat or dirt, if present, will then be taken into account when selecting components and designing adequate protection. If the factory air supply is unsatisfactory, the designer must be prepared to resist firmly pressure from ignorant, accountancy-oriented management who want the machine to start producing

before air supply deficiencies are rectified. The designer must explain clearly and firmly that the machine *will not work* unless it has a satisfactory air supply. It must be explained that contaminated air will entail the complete dismantling of all pneumatic components, cleaning and repiping before the machine is ready to work again. It must also be explained that insufficient volume or inconsistent pressure will create inconsistent power unit speeds, timing and control signals with a direct detrimental effect on the quality of the products coming from the machine. As an engineer, the designer must maintain and protect his or her professional reputation and integrity under such circumstances.

3. The efficiency of the engineering maintenance section, which, in turn, relies on:
   (a) ability and training of engineering staff in respect to systems, methods and techniques employed in the make-up of the unit.
   (b) the workshop facilities provided for the engineering section to carry out the type of maintenance required by the unit. This in itself is important if heavy losses of production time are to be avoided. There is much wasted time in many a stoppage simply because the inexpensive special tool for that pneumatic equipment was not in the workshop. False economy in this respect results in untold waste of time and, often, components damaged by improvised methods and tools.
4. The degree of communication and mutual respect which exists in the factory between management, operators and engineers.
5. The degree of simplicity which has been retained throughout all methods, arrangements and plant employed in the factory as a whole.

All of these aspects form the necessary ingredients for the successful recipe leading to productivity with quality control. The underlying and most difficult factor through it all is simplicity.

As may be seen from the progressive steps described in the overall build-up of a successful circuit for a cycle of operations, the circuit design is not a simple matter of sitting down with a paper and pencil and drawing up a pattern of symbols. The successful circuit relies very greatly on the personal involvement of the designer of the circuit. Information obtained through an intermediary is rarely an effective method of giving the designer a true picture of the requirements and working environment of the proposed production unit. Some details which may seem unimportant to the untrained eye may be of vital importance to the success of the venture. Without a true picture of the situation to work to, the designer will usually produce something which undergoes many an on-site modification at considerably greater expense before it will perform satisfactorily. This is often the cause of unexpected expenditure, far in excess of the budget figure set for the project. Close, first-hand examination of the situation by the designer is usually the least costly method in the end. The cheaper intermediary can prove a misleading and expensive misfit in the project as a whole. There can be no wholly satisfactory substitute for personal involvement by the designer.

Providing signals to initiate steps in a process

Before the alternative methods of dealing with the disposal of the unwanted pneumatic signal in a sequential operation are examined in detail, it will be worth considering for a moment the difference between the two main lines of approach to the problem which have evolved over the years.

Time estimate approach

The time estimate concept underlies the operation of many systems offered by manufacturers of instrument and control equipment. These systems take many different physical forms. They range from the early geared motor, cams and revolving drums to slow moving, endless paper belts, on which holes have been punched, combined with low pressure jet sensors. The cams operate microvalves and the holes in the paper operate jet sensors. Signals from these are of controlled duration and are produced at what are considered by the project engineer to be appropriate intervals. Once the device has been adjusted for the intervals estimated to be appropriate, they remain inflexible until a readjustment is made.

Signals from the cam-operated microvalves or jet sensors incorporated in the systems using revolving drums or endless paper belts are all on/off types of signals. In the off position, the sensor exhausts the pressure signal it provided in the on position. When the systems are adjusted before the machine controlled is started up, care is taken to ensure that a signal on one pressure end of a double pressure-operated relay valve is removed and exhausted before a pressure signal is applied to the opposite pressure end of that particular relay valve. Thus the problem normally known as the "disposal of the unwanted signal" is dealt with at the time of setting up cams and jets.

However, it must be remembered that systems of the nature described above are based on an estimate of the length of time required for the particular cycle of operation. This estimate will be the sum of the estimates of each individual time requirement for the progressive steps in the sequential operation of the cycle. In estimating the time required for each stage of the process, a margin of error must always be allowed to cover contingencies. This margin over the actual time taken to complete any particular stage may be quite small—perhaps just a few tens of milliseconds.

Although the individual margins allowed to cover variables in performance at each stage appear to be small, when reviewed accumulatively they will be seen as significant factors with a substantial influence on the final productive output of the plant. A practical example illustrates clearly the difference in productive capacity if a time estimate has to be made of each stage, with its essential allowance for error, as compared with any method which can eliminate the need for such an allowance.

The case of a high speed production unit stamping, shaping and drilling metal components will serve as an example. With eight stages in the operating cycle, it was decided that a safe margin of error of 30 milliseconds per stage would be reasonable. Estimates for the actual performance of the eight stages totalled 2 seconds. Adding the margin to cover possible errors or unforeseen variables, the overall cycle was estimated to take 2.24 seconds (2 seconds plus eight periods of 30 milliseconds). In actual fact, it was discovered that by sensing the completion of each stage and using a signal derived from the sensing of the completion of one stage to initiate the next, the cycle time was reduced to an average of 2 seconds overall. Reducing cycling time from 2.24 to 2 seconds resulted in an increase in production of 19.99 per cent.

A still more striking example of the possible effects on production of the time estimate method, when applied to the control process, is that of a factory making concrete products. The designers of the control system used a series of timers to bring in each stage of the process. These were set to operate on the time required for each stage to reach completion—mixing, forming, and the various processes involved in the curing of the final product. The cycle of operation per batch was estimated to require a total of fourteen hours. Since this system had been accepted by manufacturers in a number of different countries without question, the manufacturer here did nothing to alter the system for twelve months. However, by the end of twelve months, observation and inquiries as to types of sensors available convinced the engineer of the plant that the actual time required for each stage was considerably shorter than the estimates. Accordingly, sensors capable of sensing the change of state and completion of each stage in the process were installed. As each sensor signalled the satisfactory completion of the stage, its signal was used to initiate the next stage. The cycle of operation there now takes eight hours—not fourteen.

Feedback method

The feedback method is that in which the signal indicating the completion of a stage in the cycle of operation is used to initiate the next stage of the cycle.

9

The cascade system of sequential control is a good example of such a method. There is no time lost in waiting for an estimated time to elapse before the signal for the next stage is applied.

An understanding of the two basic concepts—time estimate and feedback—will enable the engineer to recognise them in their various guises when evaluating the commonly used sequential control alternatives described in the chapters which follow.

CETOP: The internationally accepted pneumatic symbols

A prerequisite to any discussion on circuit design is familiarity with the internationally accepted pneumatic symbols known as CETOP. These symbols are accepted by the International Standardisation Organisation (ISO). When using an international language of

this nature, it is essential that there should be no deviation from the symbols themselves. Deviation or modification of a symbol, no matter how desirable it may seem to be, can lead only to confusion and defeat of the whole purpose of the system.

The use of symbols to show the type, function and relationships of components of a system has long been accepted as an essential tool in the design, construction and maintenance of pneumatically operated machinery. In the design stage, practice has proved that, if a problem can be dealt with symbolically, the mind can be freed from questions of physical design of cylinders and valves used, and a solution obtained and circuit designed in a very rapid and sure fashion.

Accordingly, the commonly used pneumatic symbols, as promulgated and devised by CETOP are provided on pages 11–12 as a quick reference in case of any doubt arising in the mind of the circuit designer.

The more commonly used CETOP symbols

Working flow line supply and return

Pilot control line

Drain or bleed line

Electrical line

Pipeline junction

Crossed not connected pipelines

Compressed air source

Compressed air source simplified version

Electric motor

Air compressor

Rotary air motor uni-directional

Rotary air motor bi-directional

Rotary bi-directional air motor with variable capacity

Oscillating linear air motor

Semi-rotary air actuator

Single-acting cylinder

Double-acting non-cushioned cylinder

Double-acting cylinder with adjustable cushioning

Air on oil pressure dash-pot

Two-port valve

Three-port valve two positions

Four-port valve two positions

Five-port valve two positions

Five-port valve three positions

Non-return valve

Spring-loaded non-return valve

Tee or shuttle valve

Quick exhaust valve

Pressure relief safety valve

Adjustable pressure regulator with relief port

Adjustable pressure regulator with relief port and gauge

Adjustable flow restrictor or throttle valve

Uni-directional flow regulator		Flow meter	
Shut-off valve		Pressure electric switch	
Exhaust port untapped		Push-button valve operation	
Exhaust port tapped or threaded to take connections		Lever operation	
Silencer		Foot-pedal operation	
Filter		Plunger mechanical operation	
Filter with manual drain		Spring operation	
Filter with automatic drain		Roller operation	
Drier using dehydrating agent		Roller operation, one direction only	
Lubricator		Solenoid operation	
Service unit, comprising filter, pressure regulator, lubricator		Pressure operation	
Simplified version of service unit		Pressure relief operation	
Pressure gauge		Differential pressure operation, larger and smaller areas for applied pressure	

Chapter 2

Cascade system of pneumatic sequential control
Fundamental structure

The fundamental structure, pattern of thought and methods of the cascade concept are dealt with in considerable detail in *Practical Pneumatics*. There, the object was to provide an understanding of the system for the practical engineer rather than to give the total picture from a circuit design point of view. However, recapitulation here will not go amiss. If various systems are to be compared, it will be desirable this time to present the complete picture of both the cascade fundamental structure and the complementary circuits (see Chap. 3). This will establish the cascade as a method of dealing with almost all sequences the designer will be called upon to deal with in applications where the desirable control medium is pneumatic.

It is always difficult to draw a hard and fast line which will define the limits of those areas in which the cascade or, for that matter, any other pneumatic method is preferable to some other control medium. Much depends on the working environment itself together with the type of process. There are many production units working in factories using up to six and seven group cascade systems involving the control of sequences of forty or more events. There are many more processes controlled entirely by the cascade pneumatic system in which the process has a number of interlocked subcircuited units. Each of the subcircuits may have any number of cascaded groups, from two up to five or six. In these cases, both the working conditions and the type of process demand pneumatic controls as being the most suitable from cost, protection, maintenance, reliability and productivity aspects.

When carried to the extreme, the control system can in itself become too bulky and costly, although with the miniaturisation of pneumatic control components these objections are now largely academic when relating the system to the project in its practical context.

Primary purpose of the cascade system

The concept of the cascade system for pneumatics was first proposed and outlined in a paper published in 1954 by A.M. Salek. Its purpose was twofold:

1. It offered a system of disposing of the "unwanted" signal. This is a problem always associated with pneumatic control. When a pressure signal was about to be applied to the pressure end of a relay valve, it would often be found that a pressure signal from an earlier movement still remained on the opposite end of the relay. Before any further action could take place, the first signal had to be removed by some means or other.

2. It offered a method of ensuring that no movement or event in a cycle of operation could take place until the preceding movement had been completed. In other words, the feedback signal indicating the successful completion of a movement is also the signal used to initiate the next event in the cycle. With no time lost between movements, it was a major step forward in increased productivity of a cycle of operation.

The cascade system not only fulfils both these objectives, but it also still remains the simplest, most economical pneumatic control system over a wide range of operations, from simple to remarkably complex. Its simplicity

and logical structure not only make it easy for the average maintenance engineer to understand with minimal instruction, but also provide a system which renders diagnosis of an unscheduled stoppage easy and quick. In turn, this reduces unproductive down time for the machinery concerned to a low level.

The unwanted or opposed signal

Before examining the steps taken to draw up a cascaded circuit diagram, it will be worthwhile looking at the primary reason for the cascade. This is the problem of the "unwanted" signal—the signal which must be removed before another signal may be applied to the opposite pressure end of a relay valve. Figures 2.1 and 2.2 show just how this state of opposed signals can occur in a simple sequence.

In Figure 2.1, it can be seen that if the start switch is switched to the "on" position, a pressure signal from the sensor marked $a-$ will pass through the switch to the pressure end of A cylinder's relay. This relay, however, already has pressure on the opposite end of the relay spool from the sensor marked $b-$ which is still depressed. Since the ends of the spool in the relay valve are of equal diameter and area, and the pressure from both $a-$ and $b-$ are equal, there can be no movement of the spool in

Fig. 2.1 *Opposing signals on A's relay*

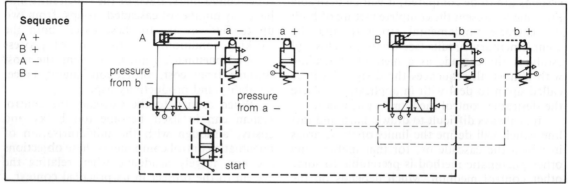

Sequence	
A +	
B +	
A −	
B −	

pressure from b −

pressure from a −

start

A   a −   a +

B   b −   b +

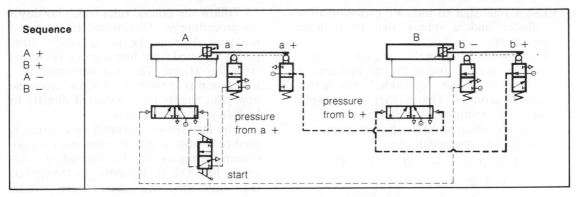

Fig. 2.2 *Opposing signals on B's relay*

response to the signal from the start switch until the pressure from $b-$ sensor is removed.

In Figure 2.2, the circuit has been drawn to show the situation which would occur if by some means or other A cylinder had extended so that the sensor marked $a+$ were depressed to allow pressure to be applied to the pressure end of B cylinder's relay. In turn, it is imagined that by some means or other this had caused B cylinder to extend and depress the sensor marked $b+$. Pressure then from $b+$ is shown on the diagram as being applied to the opposite end of B's relay. Now, as the diagram shows, there is pressure on one end of B's relay from $a+$ and pressure on the other end from $b+$. The intended sequence required B to retract at that stage but, with pressure on each of its equal ends, the spool in B's relay cannot move. The only way in which it will move is to remove the pressure from $a+$. This, then, is the problem which the cascade system never fails to overcome.

It should be noted that not all types of sequence will produce the problem of opposed signals when the circuit is drawn up. Those that do not encounter this problem are commonly termed "simple circuits". Those sequences which require only a simple circuit are easily recognised after the sequence has been written down, as will be seen as the consecutive steps which are taken in designing a circuit are described.

The remainder of this chapter has two distinct parts:

1. a precis of the cascade format for pneumatic sequential control for those who are thoroughly familiar with and using the system in their day-to-day design.
2. a step-by-step description of the cascade method in practice for those whose normal activities include so many areas that pneumatic circuitry can claim only limited attention. For such situations, the step-by-step build-up provides a work pattern which can be used and directly related to any practical problem as it occurs.

A precis of the cascade format

1. Identify each movement in the sequence of operation, together with the order in which they occur. *Note:* The term "movement" used here can mean any event in the process such as the turning on and off of a control valve in a fluid pipeline, etc.
2. Allot a letter to represent each movement and write the sequence down, e.g.

$$A + B + C + B - A - C -$$

Use a plus sign to indicate extension of a cylinder and a minus sign to indicate retraction.

3. Examine the letters in their sequence. If they do not present the order of appearance known as "direct repetition", the letters must be grouped. That is, they must be split into the minimum possible number of groups in which no letter appears more than once in any one group, e.g.

$$A + B + C + / B - A - C -$$

Where one letter appears once in a cycle in which all the other letters appear twice, followed by the same pattern in the next cycle, refer to the method described in Chapter 3 which deals with this. Or, where a letter appears to indicate that a movement oscillates several times in a cycle in which all the others appear only twice, again refer to Chapter 3 under the section dealing with this type of sequence, e.g.

$$A + B + C + C - A -,$$
$$A + B - C + C - A -$$
or
$$A + B + B - B + B - C + C - A -$$

4. Having grouped the letters, number the groups using Roman numerals, e.g.

$$\overset{\text{I}}{A + B + C +} \overset{\text{II}}{/ B - A - C -}$$

5. Draw the required components:
   (a) one cylinder for each letter.
   (b) one five-port relay valve for each double-acting cylinder.
   (c) two sensors for each cylinder.
   (d) one start switch.
   (e) five-port logic valves for the pilot supply system:
      (i) one such for a two-group system pilot supply configuration;
      (ii) two such for a three-group pilot supply configuration;
      (iii) one such valve for each group of four or more, using the pilot supply configuration for four or more groups.

6. Connect up the components using broken lines for signal lines, solid lines for supply lines.

Follow the general rules which lay down the procedure as: All movements in a group are piloted from the group supply line, group I piloted from line I, group II piloted from line II, etc. The first movement in a new group is initiated by a signal from the new pilot supply line connected directly to its five-port control relay.

The last sensor activated in a group is used to change over the group pilot supply system, bringing in the succeeding pilot supply line. Often, depending on the type of logic valve used, this selector sensor will be supplied from a main air source. The start switch is connected up merely to interrupt the first signal in the cycle of operation which initiates the first movement in the cycle.

7. Complete the circuit diagram by filling in all relevant information such as cylinder letters, functions, component identification number, sequence of operation, etc.

Practical consecutive steps in circuit design

Simple circuits

Write the sequence down

After the sequence of events has been positively identified, write down each event in the order in which it happens in the cycle of operation. As a working example, imagine the sequence to be comprised of four movements. The first movement will be written down as A. If it is a cylinder extending or, perhaps, a valve being turned on, it will be written as $A +$. If it is a cylinder retracting or a valve being turned off, it will be written as $A -$.

Going through the sequence of our working example it is found that the sequence when written down proves to be:

$$A + B - C + D - A - B + C - D +$$

The letters are then examined as to the nature of the order in which they appear.

In this case the letters fall clearly into two groups in which the order of the letters in the second group is identical to the order in which they appear in the first, namely, *A B C D*, then *A B C D*. In this context such a sequence is usually referred to as "direct repetition".

Whenever the sequence is identified as direct repetition—the order of letters repeated, regardless of the plus or minus—the circuit required will follow the pattern known as a simple circuit. This pattern still ensures that no movement will follow another movement until the preceding movement is complete. At the same time, the sensor detecting the completion of the movement will provide a signal which will initiate the next movement with the least possible delay. From a productivity aspect, this will ensure that the limiting factors in the time taken by the cycle of operation will be confined, in a well-planned sequence, to the sum of the times taken by the individual productive movements only, plus the few milliseconds required for each signal and valve changeover.

Draw the prime movers

Before drawing the prime movers, as shown in Figure 2.3, write the sequence down the side of the diagram. Then draw each of the prime movers in the state at which each is in a position of rest, just before the cycle of operation starts.

In the working example, as can be seen on Figure 2.3, there are four cylinders: *A* is a double-acting cushioned cylinder retracted at rest; *B* is a double-acting cushioned cylinder extended at rest; *C* is a single-acting cylinder retracted at rest; and *D* is a double-acting cushioned cylinder extended at rest. Draw these and give each its identifying letter.

Fig. 2.3 *Step 1: Draw prime movers and sequence*

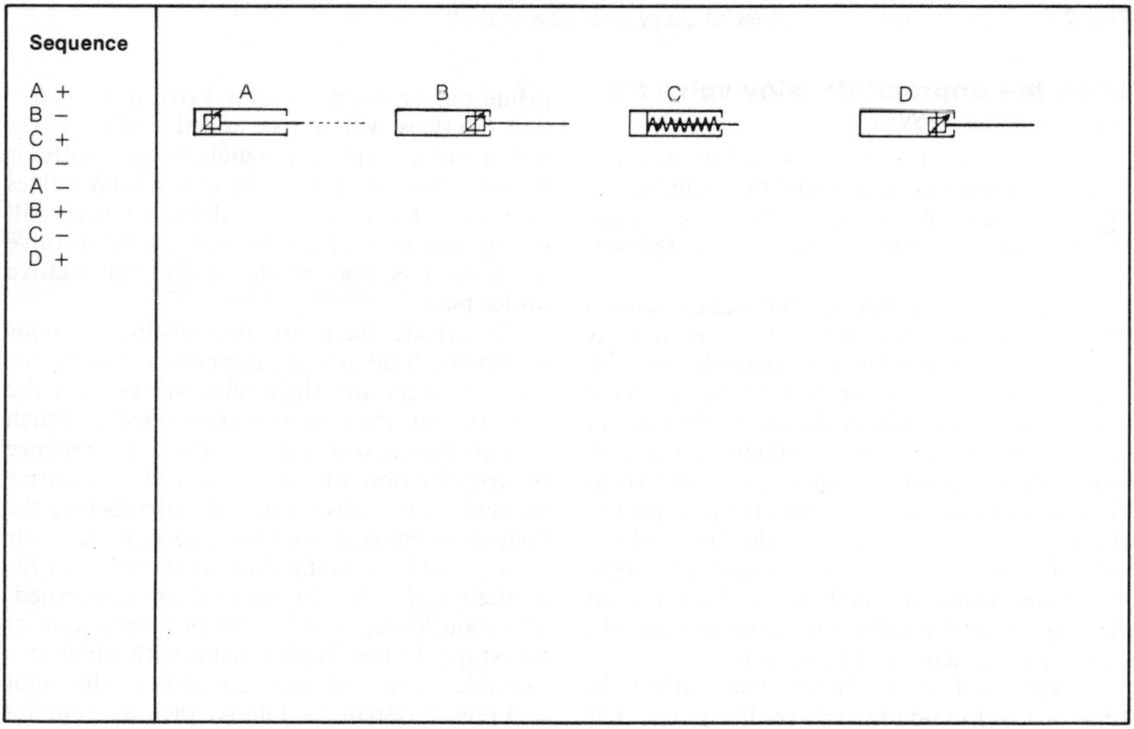

Sequence

A +
B −
C +
D −
A −
B +
C −
D +

Sequence				
A +				
B –				
C +				
D –				
A –				
B +				
C –				
D +				

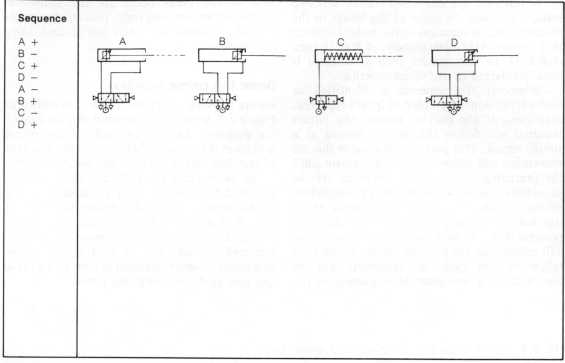

Fig. 2.4 *Step 2: Add relay valves to complete power units*

Draw the appropriate relay valve for each prime mover

Draw in the connecting lines, valves to cylinders. Note on Figure 2.4 how the cylinders *A*, *B* and *D* require five-port valves and *C*, the single-acting cylinder, requires a three-port relay valve.

Draw the main air symbol—circle with a central dot—at each inlet port of each relay valve. Draw the exhaust port symbols. Note on Figure 2.4 how the connections to the valves are drawn to the box which shows the flow of air through the valve when the cylinders are at rest. With the retracted cylinders, air is shown as flowing through the valve from the inlet port to the outlet port connected to the front of the cylinder. With those cylinders extended at rest, the connections are such as to show the air flowing from the valve inlet port through the outlet to the rear of the cylinder.

Figure 2.4 now shows what might be regarded as the power units of the system. The prime mover needs its relay valve to be effective. To these will now be added a pilot system which will use pressure signals to push with air pressure the internal spools of the relay valves one way or the other. In doing so they will change the flow of air through the relay valve from its inlet port to the second, alternative outlet port.

In effect, there are two distinct systems within the total system: the power system, the prime movers and their relay valves, and the pilot system, the sensors and logic valves which may be required to gain a particular sequence or combination of predetermined alternative sequences. It is always worth remembering the difference between the two systems and, so far as is possible, keeping them separate in so far as their respective air supplies are concerned. The significance of this will not be apparent at this stage. In the simple circuit, with which this example is immediately concerned, the pilot and power system both derive their air supplies

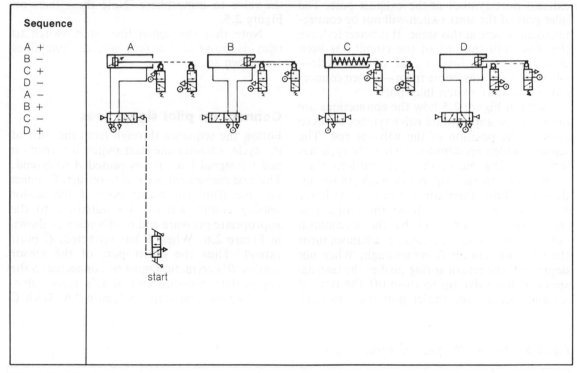

Fig. 2.5 *Step 3: Add pilot sensors and start switch*

from a common source. Later, however, the benefits to be derived from cultivating a mental awareness of the differences between the two will become obvious.

Draw the pilot components

Add to the drawing the pilot components which will control the sequence of operation of the power units already drawn. By referring to Figure 2.5, it will be seen that the pilot system in this case is comprised of a sensing device at each end of the stroke of each prime mover. In addition a start switch is required. Draw this and mark it "start".

The sensors shown represent roller-operated three-port valves. When depressed by the arrival of the cams attached to the movements made by the cylinders, these valves will, as each is depressed in turn signifying the completion of the respective stroke, pass the signals indicating

stroke completion and initiate the next movements.

The start switch is a switch-operated three-port valve which will, when required, interrupt the signal initiating the first movement of the cycle so that the cycle of operation may be stopped at will by changing the switch to the off position. After the switch has been turned off, the cycle will continue through to the completion of the last movement. The signal indicating completion and initiating the start of the next cycle will then find it cannot get past the start switch and so the next cycle cannot start.

Having drawn in the sensors, since this is a simple circuit, the main air supply will be drawn at each inlet port of each sensor as shown in Figure 2.5. Also the exhaust port symbol will be drawn then at each exhaust port.

The signal line from start switch to the relay valve for the first movement of the cycle will be drawn as shown in Figure 2.5, together with the exhaust port symbol at the exhaust port. The inlet port of the start switch will not be connected to anywhere at this stage. It is better to leave this line until the rest of the circuit has been drawn and the final signal of the cycle—wherever that may come from—is then connected to the start switch inlet port.

Note in Figure 2.5 how the connections are drawn to the appropriate valve symbol boxes to indicate the position of the valve at rest. The sensors, which are depressed when the cycle has stopped at the end of the cycle, all have their connections to the top boxes while those not depressed have their connections to the lower boxes. The box always shows the position of the valve when operated by the mechanism adjacent to it. Thus, the roller mechanism turns the valve on and air flows through. When not depressed, the return spring pushes the internal spool of the valve up to close off the flow of air and exhaust the outlet port back through

the valve to atmosphere. Note these facts on Figure 2.5.

Note that the signal line, start switch to relay, is drawn as a broken line; all signal lines are drawn so.

Connect up pilot signal lines

Follow the sequence through from the start of the cycle. Imagine the start switch is turned on and the signal from it has caused A to extend. The next movement is for B to retract. Connect the line from the outlet port of the sensor sensing completion of A's outstroke to the appropriate pressure end of B's relay as shown in Figure 2.6. When B has retracted, C must extend. Thus the outlet port of the sensor sensing B's retraction must be connected to the appropriate pressure end of C's relay valve. This line may be seen on Figure 2.6. With C

Fig. 2.6 *Step 4: Connect up signal lines*

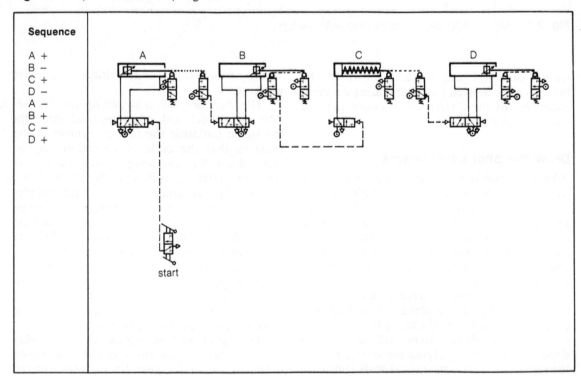

Sequence
A +
B −
C +
D −
A −
B +
C −
D +

start

extended, the sensor depressed at C's outstroke must send a signal to D to retract D. Thus, it will be seen on Figure 2.6 that a line from the outlet port of C's outstroke sensor must be connected to the appropriate pressure end of D's relay valve.

As is seen on Figure 2.6, the cycle has progressed to the half-way mark. To complete the cycle, connect up the lines from the sensors to their respective relays as determined by the sequence written on the side of the diagram, following through movement by movement.

Figure 2.7 shows the signal lines all connected up and, as a reminder that any circuit diagram will have a practical purpose, the symbol for a service unit—filter, pressure reducer and lubricator—has been drawn to complete the circuit.

Summing up, a simple circuit will suffice if, when written down, the sequence displays a pattern termed "direct repetition". This applies to the pattern in which the letters fall into two groups, with the same sequence of letters. In such cases, all that is required to attain a positive control of the sequence is to provide the prime movers and their respective relay valves, two sensors for each prime mover to determine each completion of the instroke and outstroke of each prime mover, and a start switch. The relay valves controlling the prime movers and the sensors will all be supplied directly from the main air source—clean, dry and, in most cases, lubricated air. If the sequence does not display the direct repetition pattern, the circuit design will require to follow the cascade method.

Fig. 2.7 *Add service units to complete circuit*

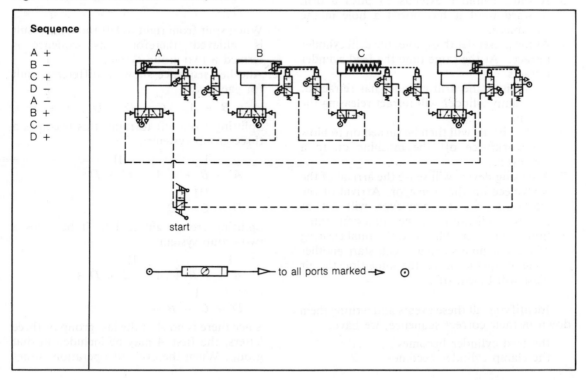

Practical consecutive steps in circuit design

Cascaded circuits

Write the sequence down

After the sequence of events has been identified, write it down in the manner described for simple circuits. That is, allot a letter to each event, starting with the first as the letter A, together with a plus sign for an outstroke of a cylinder or the turning on of a valve, heat, water, etc. A minus sign will signify the reverse—an instroke, turning off, etc.

As a working example of such an exercise from the start, take the case of a machine which has been planned to work in the following manner:

1. A cylinder extends to push a workpiece into position.
2. Another cylinder extends to clamp the workpiece in position.
3. A third cylinder extends to push a drill forward until it has bored a hole in the workpiece.
4. At the preset depth of hole, the drill cylinder retracts. At the same time the feed cylinder retracts.
5. As soon as the drill cylinder has retracted, the clamp cylinder will retract releasing the clamp.
6. A blast of air will then be turned on to blow the workpiece off the machine on to a conveyor.
7. A sensing device will sense the arrival of the workpiece on the conveyor. Arrival of the workpiece on the conveyor will act as positive indication of the successful functioning of the air blast and the final clearing of the machine so that it can start another cycle of operation. At the same time the air blast will be cut off.

Identifying all these events and writing them down in their correct sequence, we have:

the feed cylinder becomes	A
the clamp cylinder becomes	B
the drill cylinder becomes	C
the air blast becomes	D

The sequence when written down appears as:

$$A + B + C + C - B - D + D -$$
$$ A -$$

Group the letters

Number the groups, using Roman numerals. The letters, written in their sequence, will then be examined to make sure that the pattern of direct repetition is not shown. If it is not, the letters are divided, or split, into the least possible number of groups in which no letter appears more than once in any group. The letters should be split into groups working first from left to right then right to left. The reason for this becomes obvious when the following examples are put to the test.

1. The working example split from left to right results in two groups:

$$\begin{array}{lc} \text{I} & \text{II} \\ A + B + C + / & C - B - D + / \\ D - & A - \end{array}$$

When split from right to left the same result is achieved; therefore this sequence is termed a two-group system.

2. Another sequence shows a different result for each way:

$$A + B + A - C + D + D - C - B -$$

Splitting from left to right, this becomes a three-group system:

$$\begin{array}{lc} \text{I} & \text{II} \\ A + B + / & A - C + D + / \\ \text{III} & \\ D - C - B - / & \end{array}$$

Splitting from right to left, it becomes a two-group system:

$$\begin{array}{lc} \text{I} & \text{II} \\ A + / & B + A - C + D + / \\ \text{I} & \\ D - C - B - & \end{array}$$

Since there is no A in the last group of three letters, the first A may be included in that group. When the cycle of operation comes into effect, the start switch will be used to interrupt the signal from the sensor sensing the completion of *B*'s instroke so that the continuous cycling will stop at that point.

The significance of the grouping is that at some part of each group there will be a potential problem of opposed signals. The sequence cannot be altered to release a sensor delivering a signal which will find itself opposed by another signal already at the opposite pressure end of the relay which must be changed over by the second, later signal. The alternative is to exhaust the air pressure supplying the first sensor so that pressure from the second signal will be effective. With the first signal removed, the relay will change over. Thus, each group will have its own air supply for the sensors operating in the group of movements. If group I is operating, all other pilot pressure supply lines will be exhausted to

atmosphere. Only one pilot supply may be on at any time.

Proceeding, then, with the working example, having found that the cycle is a two-group system, the next step is to start the actual drawing of the circuit.

Draw the prime movers for each event and letter them

From Figure 2.8 it will be seen that the prime movers—cylinders *A*, *B* and *C*—have each been given their respective five-port relay valves so that, in effect, valve and cylinder become a power unit as distinct from any part of the pilot system. The air blast *D* has been given a two-port poppet valve to perform its function and, for the time being, a three-port relay valve has been allotted to operate the poppet valve. Each has been given its designatory letter in the sequence and the sequence itself has been

Fig. 2.8 *Step 1: Draw prime movers and their relay valves*

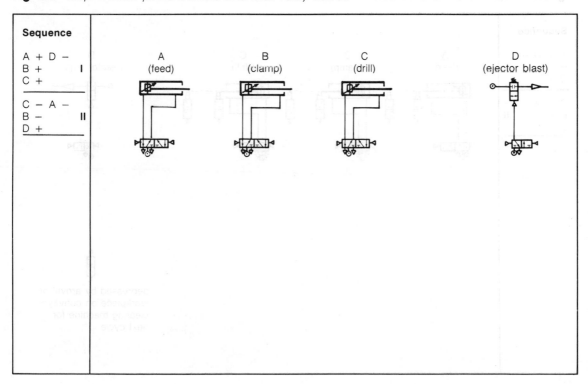

written down on the side of the paper on which the diagram is now being drawn. As an aid to those who will use the diagram later, both for installation and maintenance, the function of each of the prime movers is shown in brackets. All necessary connections have been drawn and the main air supply symbol drawn at appropriate supply ports, together with appropriate exhaust port symbols.

With the power units now on paper, the next step is to start drawing the pilot system which will make them perform in the sequence shown.

Draw the sensors

Note in Figure 2.9 that the sensors sensing mechanical movement are shown as related to the cylinder powering the movement. For example, the mechanical device which is moved by the extension of cylinder A's piston rod will have some sort of cam which will depress a sensor at the extremity of the movement, or

when cylinder A has completed its outstroke. The sensor, when depressed by the cam, will give a signal to initiate the next movement in the cycle. In the diagram, the sensor referred to here is shown as being positioned at the end of A's outstroke. The sensor sensing the arrival of the workpiece must have some brief reference alongside to indicate its purpose in the process.

Draw in the start switch and mark it

Figure 2.10 shows the diagram at this stage, ready for addition of the pilot supply valves.

Draw the pilot supply valves and the pilot supply lines

As this is a two-group system, two pilot lines will be required—one for each group. This will be obtained from one small, five-port double pressure-operated valve.

Figure 2.11 shows the addition of the pilot supply valve and the supply lines which, for

Fig. 2.9 *Step 2: Draw pilot sensors*

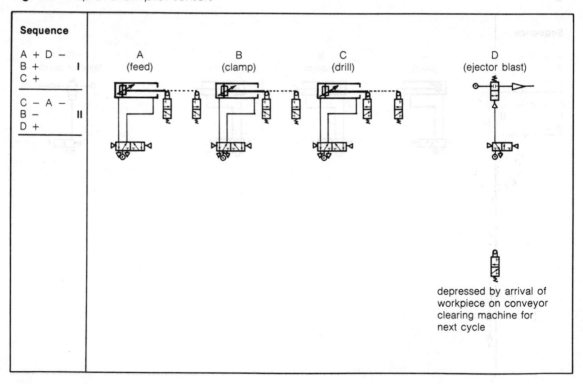

Sequence

A + D −
B +   I
C +

C − A −
B −   II
D +

A (feed)   B (clamp)   C (drill)   D (ejector blast)

depressed by arrival of workpiece on conveyor clearing machine for next cycle

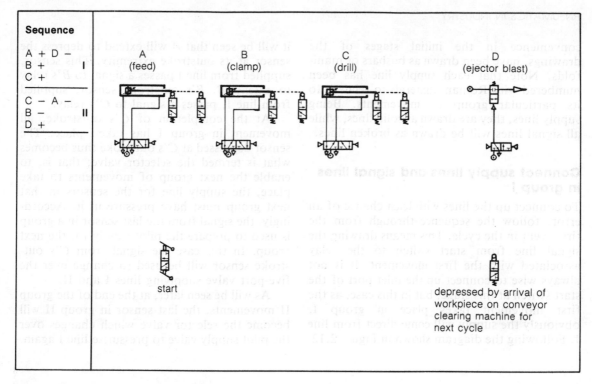

Sequence

A + D −
B +   I
C +

C − A −
B −   II
D +

A
(feed)

B
(clamp)

C
(drill)

D
(ejector blast)

start

depressed by arrival of
workpiece on conveyor
clearing machine for
next cycle

Fig. 2.10 *Step 3: Add start switch*

Fig. 2.11 *Step 4: Add pilot supply valves*

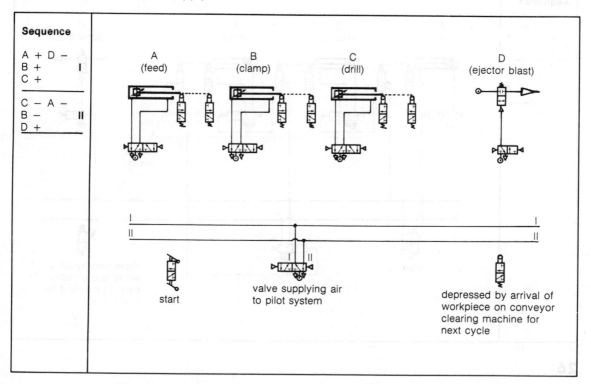

Sequence

A + D −
B +   I
C +

C − A −
B −   II
D +

A
(feed)

B
(clamp)

C
(drill)

D
(ejector blast)

I

II

start

valve supplying air
to pilot system

depressed by arrival of
workpiece on conveyor
clearing machine for
next cycle

convenience in the initial stages of the drawings, have been drawn as busbars or manifolds. Note that each supply line has been numbered with Roman numerals to relate it to its particular group of movements. Being supply lines, they are drawn as solid lines, while all signal lines will be drawn as broken lines.

Connect supply lines and signal lines in group I

To connect up the lines with least chance of an error, follow the sequence through from the first event in the cycle. This means drawing the signal line from start switch to the relay associated with the first movement. It is not always wise to connect up the inlet port of the start switch at this stage, but in this case, as the first movement takes place in group I, obviously the supply will come direct from line I. Following the diagram shown in Figure 2.12,

it will be seen that *A* will extend to depress the sensor at its outstroke extremity. This sensor, supplied from line I passes a signal to *B*'s relay to extend *B*. *B*'s outstroke sensor, supplied from line I, passes a signal to *C*'s relay.

At the completion of *C*'s outstroke, all movement in group I has taken place. The sensor depressed at *C*'s outstroke thus becomes what is termed the selector valve; that is, to enable the next group of movements to take place, the supply line for the sensors in that next group must have pressure in it. Accordingly, the signal from the last sensor in a group is used to prepare the pilot supply for the next group. In this case the signal from *C*'s outstroke sensor will be used to change over the five-port valve supplying lines I and II.

As will be seen later, at the end of the group II movements, the last sensor in group II will become the selector valve which changes over the pilot supply valve to pressurise line I again.

Fig. 2.12 *Step 5: Connect group I movements' supply and signal lines*

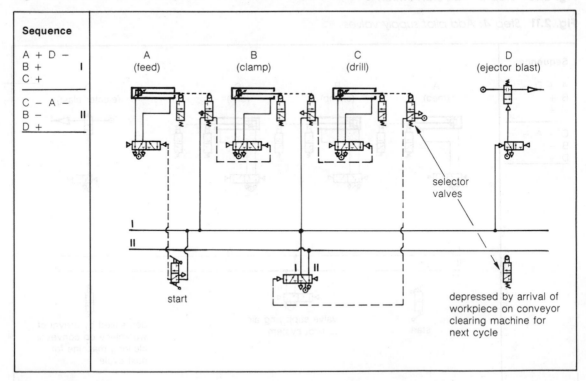

In this case, the last sensor to be operated in group II is that which detects the arrival of the workpiece on the conveyor. Its signal will be used to change over from line II to line I.

Connect supply lines and signal lines in group II

The movements in group II are followed through in accordance with the sequence written on the side of the diagram, and the supply and signal lines drawn from sensor to sensor until the end of the cycle. Note in Figure 2.13 that although A and D are shown in the sequence as acting together, only the signal to A's relay passes through the start switch. Line I is drawn direct to D's relay. This is because, when the machine is stopped by switching off the start switch, the signal for A's first movement will not be delivered. On the other hand, the ejection air blast which has fulfilled its

function does not need to be kept in the "on" position any longer. If its "off" signal passed through the start switch when the switch was turned off, all movement would cease but the air blast would continue to blow.

Note that the selector valve to select line I is the sensor sensing the arrival of the workpiece on the conveyor. Note also that both selector valves are supplied at their inlet ports from the main air supply, rather than from the pilot supply. The reason for this is that some types of valve, by their design, are not reliably changed over with air coming from their own ports. In other words, using air from a valve to blow its own spool over and cut off the air which actually blew the spool across is not considered good practice when applied to all types of valves. Usually valves which cannot use what is termed a "dying signal" for such a purpose are using a seal design known as a static seal. This is referred to in some detail in *Practical Pneumatics*.

Fig. 2.13 *Step 6: Connect group II supply and signal lines*

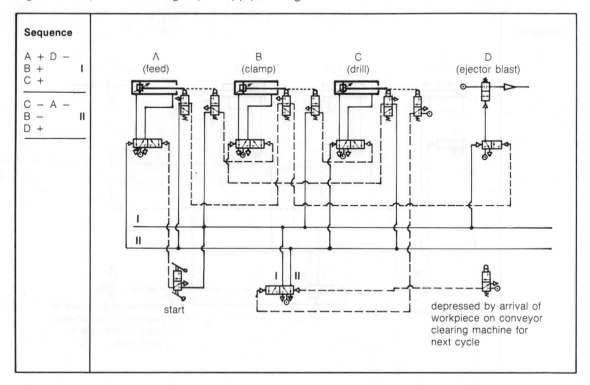

Review the completed circuit for any redundant valves

Now that the sequence is assured positively and no movement can follow another until the first has completed its function, some thought should be given to the question of tidying up the circuit. It should be looked at with a critical eye to make sure that there are no unnecessary components and also that all non-productive movements have been carried out as much as possible while productive movements were taking place. Also, where several movements may occur together, it is wise to make sure that they have all been completed before the cycle has reached the stage where they are required to move again. Note that, in the circuit shown in Figure 2.13, the signal which turns on *D* has originated from line II, passed through *A*'s instroke sensor, then through *B*'s instroke sensor. This ensures that before ejection is attempted, the clamp has been released and the

feed cylinder has retracted fully, so that when the next cycle is initiated, straight after the *D* blast has done its work, the feed cylinder will be able to feed in the next workpiece.

Reviewing the circuit, then, as it is at this stage, we may be reasonably satisfied that it fulfils our requirements as a positive and reliable means of achieving our sequence of operation. It is safe to assume that the machine will be safe from potential mechanical "foul ups".

However, there is one aspect which should raise a query. Note how such a query is resolved in the circuit shown in Figure 2.14.

The query is in respect to the *D* ejector blast. The valve actually passing the air through for the blast is an air-operated, spring-return poppet valve. This poppet valve is turned on, to provide the ejector blast, by a signal from a sensor depressed when *B* reaches the end of its instroke. Pressure for this signal comes from

Fig. 2.14 *Step 7: Eliminate unnecessary valves in final diagram*

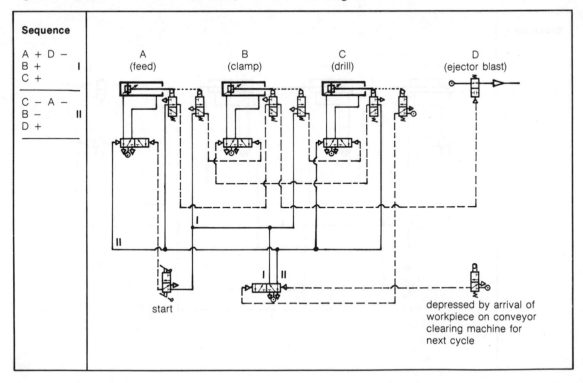

line II pilot pressure supply and is also the last signal in group II. When pilot pressure supply line I is turned on, line II is turned off and the pressure in line II exhausted to atmosphere. When pressure is removed from the pressure end of the poppet valve, the poppet valve return spring turns the poppet valve off. From consideration of these facts it becomes obvious that the three-port relay valve is redundant. The same result can be achieved by applying *B*'s instroke sensor signal directly to the pressure end of the poppet valve. The blast will remain on so long as pressure from line II is maintained through the sensor. As soon as line I is pressurised, line II will exhaust and the poppet valve close. The blast, likewise, will be turned off.

Note how the final diagram as shown in Figure 2.14 dispenses with the unnecessary three-port valve. Note also how the pilot lines have been tidied up. The circuit is now ready for the matter of reviewing each event in turn—in the case of the cylinders it is a matter of speed, thrust and materials. The air blast has to be looked at in terms of volume and pressure. All the details needed for a practical installation are reviewed carefully. The mechanics of the actual sequence, as can now be seen, are but a small part of the overall exercise.

From this example of the cascade, although only a two-group circuit, two major factors should be clear which apply to all cascaded circuits no matter how many groups may be involved.

1. The first movement of the new group is initiated by direct application of the new supply line to the relay of that first movement.
2. The sensor detecting the completion of the last event in a group becomes the selector for the supply line which will pilot the next group. As such it will be supplied from mains air.

The method of employing the cascade concept shown in this example is exactly the same with any cascaded circuit, regardless of the number of groups involved, except for one factor—the configuration of the pilot supply varies slightly for a three-group and for four or more group systems.

Pilot supply configuration and design of a three-group cascaded system

Figure 2.15 shows the detail of the configuration for the pilot supply system in a three-group system. Points to note include:

1. Lines I and II are supplied from the two outlet ports of the upper five-port valve. Of the two outlet ports in the lower five-port valve, one supplies line III while the other provides the supply for the upper valve, connected to its inlet port.
2. It is only possible for air to be pressurising one of the supply lines at any one time. The upper valve has two positions only. It will consequently pressurise line I, at which time line II will be exhausted to atmosphere, or it will pressurise line II, at which time line I will be exhausted to atmosphere. The lower valve has also only two positions. Consequently, when line III is pressurised, the supply to the upper valve and thence through to either line I or line II will be exhausted to atmosphere. In its alternative position the line to the upper valve will be pressurised, at which time line III will be exhausted to atmosphere.

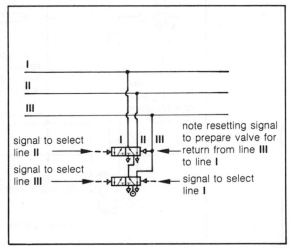

Fig. 2.15 *Three-group pilot supply configuration*

3. When drawing the configuration, always reset the valve above from line III. If this develops into a habit there can be no possibility of a ghost signal escaping into line II when a changeover from line III to line I is made.

Using this configuration makes no difference to the consecutive steps outlined in the example involving the two-group system.

Figure 2.16 provides an example of a three-group system. In this example the cycle of operation finishes with the retraction of C. The grouping calls for line I to retract C and the first movement of the next cycle is also in group I, therefore piloted from line I. Note that, in such a case, the start switch provides a means of interrupting the signal from C's instroke sensor to A's relay to initiate A's outstroke at the start of the next cycle.

Four or more group cascade systems

When a sequence has been written down and grouped, it may result in four or more groups of letters. In such cases the pilot supply system employs the configuration shown in Figure 2.17. This has become standard practice for two practical reasons. Although there are still many four-, five- and six-group systems working reliably and efficiently with a pilot configuration based on extending that for the three-group system shown in Figure 2.16, it is generally recognised that the pilot supply pressure can drop lower than is desirable when it has to pass through too many valves before it reaches its final destination. At the same time, lubrication may suffer in that, as the air stream passes through each successive valve, some of

Fig. 2.16 *Example of three-group system*

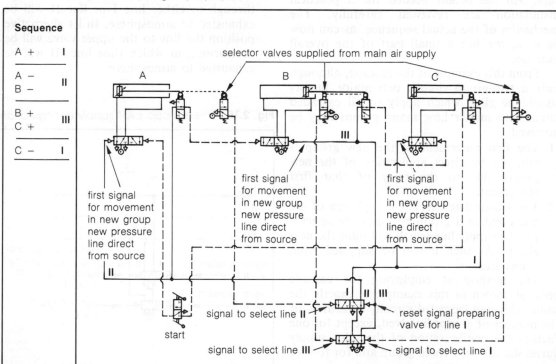

Sequence	
A +	I
A − B −	II
B + C +	III
C −	I

selector valves supplied from main air supply

A   B   C

III

first signal for movement in new group new pressure line direct from source

first signal for movement in new group new pressure line direct from source

first signal for movement in new group new pressure line direct from source

I

II

start

signal to select line II

I II III

reset signal preparing valve for line I

signal to select line III

signal to select line I

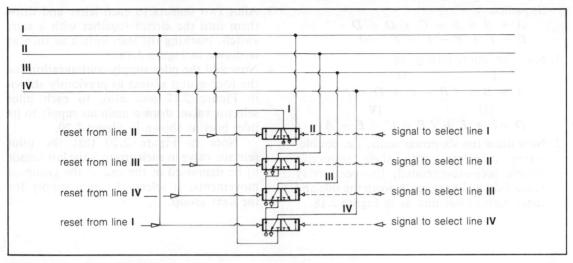

I

II

III

IV

I

reset from line II _____   signal to select line I

reset from line III _____   II   signal to select line II

reset from line IV _____   III   signal to select line III

reset from line I _____   IV   signal to select line IV

Fig. 2.17 *Four or more group pilot supply configuration*

the lubricating mist of oil, carried in suspension, is dropped off in each valve as the air finds its way through. The final valve on the end of the receiving line may find insufficient oil left to keep it adequately lubricated.

The configuration shown in Figure 2.17 ensures that these two possibilities do not apply. Since the main air supply travels through only two valves before emerging into any supply line, it can be extended to cover any number of groups simply by adding another five-port valve to the pilot supply system for each additional group. Thus a six-group system would employ six ⅛'' BSP, double pressure-operated, five-port valves arranged in the configuration shown in Figure 2.17, each valve supplying the one above in the manner shown there.

Procedure for connecting up four or more group circuits

The procedure for connecting up the circuit is along the same lines as described in detail as the procedure for drawing up the circuit shown in Figure 2.16, except for the configuration of the pilot supply system. Group line selection and valve resetting are fully outlined in Figure 2.17 for the configuration required for four or more groups.

As an example of the practical building up of a four or more group cascaded circuit, the following sequence will serve to show the steps which are generally accepted as being the most helpful and least open to error.

Sequence:

$$A + B + B - C + D + D -$$
$$E + F + F - C - E - A -$$

1. Split the letters into groups:

$$
\begin{array}{cc}
\text{I} & \text{II} \\
A + B + / & B - C + D + / \\
\text{III} & \text{IV} \\
D - E + F + / & F - C - E - A - /
\end{array}
$$

2. Now draw the six power units, i.e. double-acting cylinders, with their respective double pressure-operated, five-port, relay valves; show the sequence down the side and letter each power unit as in Figure 2.18.

3. Allot two sensors to each letter and draw them into the circuit together with a start switch, marking the start switch as such as is shown in Figure 2.19.

4. Now add the pilot supply configuration for the four-group system as previously shown in Figure 2.17 and also, to each pilot selector valve, draw a main air supply to its inlet port as shown in Figure 2.20.

Note in Figure 2.20 that the pilot selector valve in each group is the last sensor to be depressed at the end of the group of movements. It selects the pilot supply for the next group.

Fig. 2.18 *Draw sequence and power units*

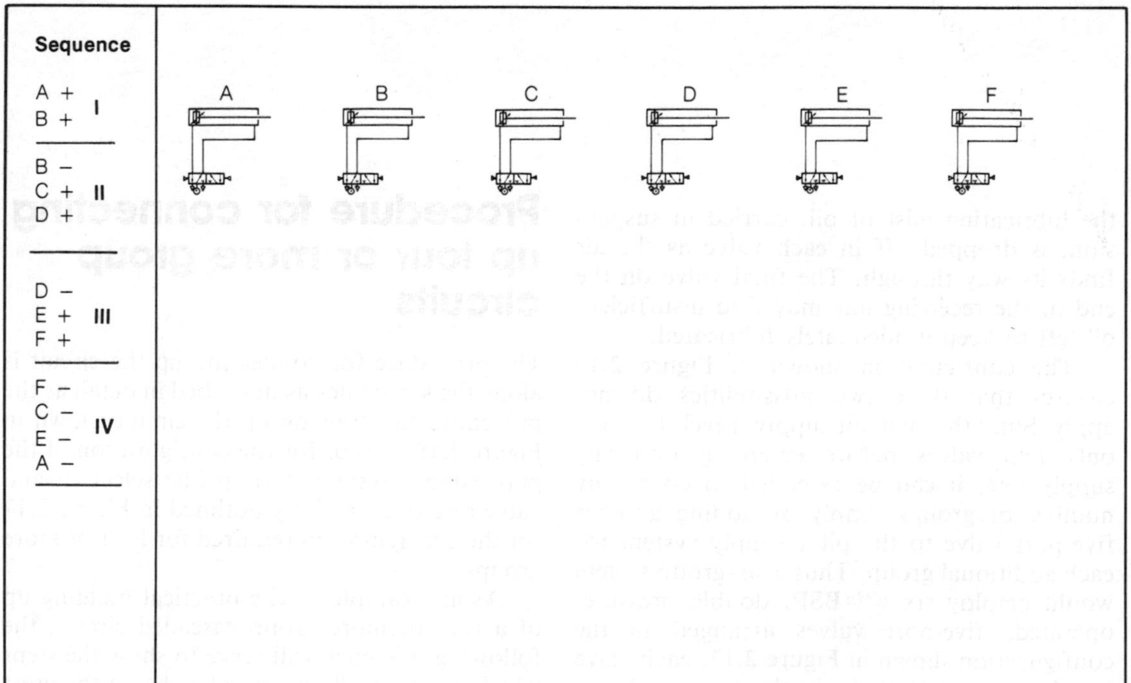

Sequence		
A +	I	
B +		
B −		
C +	II	
D +		
D −		
E +	III	
F +		
F −		
C −	IV	
E −		
A −		

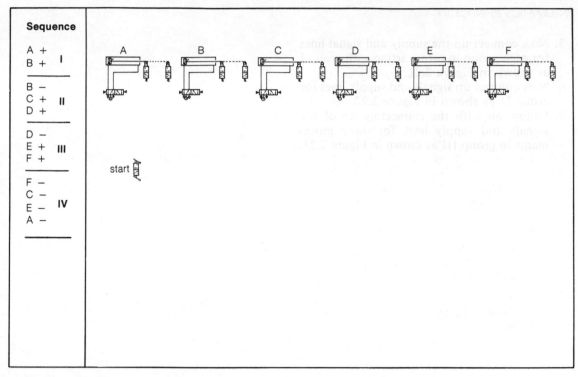

Sequence	
A +	I
B +	
B −	
C +	II
D +	
D −	
E +	III
F +	
F −	
C −	
E −	IV
A −	

start

Fig. 2.19 *Add sensors and start switch*

Fig. 2.20 *Add four-group pilot supply configuration*

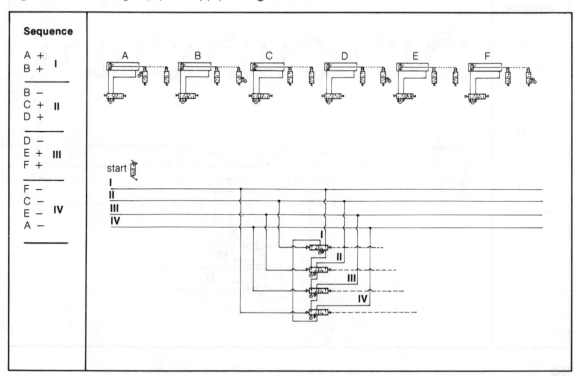

Sequence	
A +	I
B +	
B −	
C +	II
D +	
D −	
E +	III
F +	
F −	
C −	
E −	IV
A −	

start

I
II
III
IV

I
II
III
IV

5. Now connect up the supply and signal lines for each group in turn starting with group I as shown in Figure 2.21.
6. Now connect up signals and supply lines for group II as shown in Figure 2.22.
7. Follow on with the connecting up of the signals and supply lines for those movements in group III as shown in Figure 2.23.

Fig. 2.21 *Connect group I supply and signal lines*

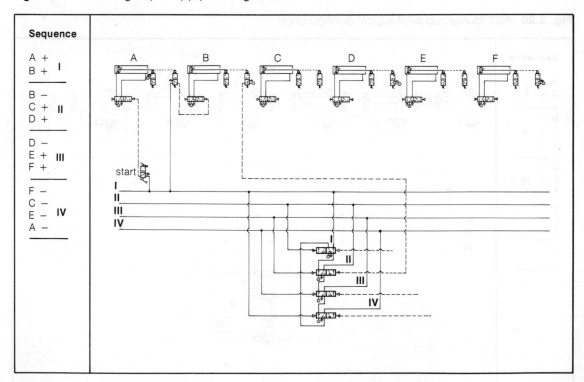

Sequence	
A + B +	I
B − C + D +	II
D − E + F +	III
F − C − E − A −	IV

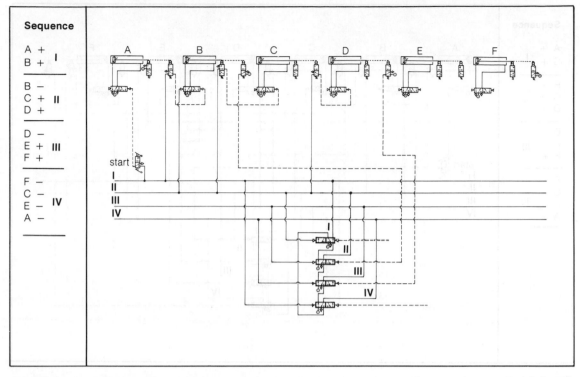

Fig. 2.22 *Connect group II supply and signal lines*

Fig. 2.23 *Connect group III supply and signal lines*

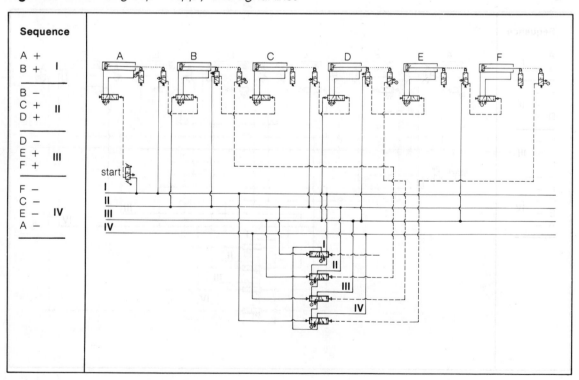

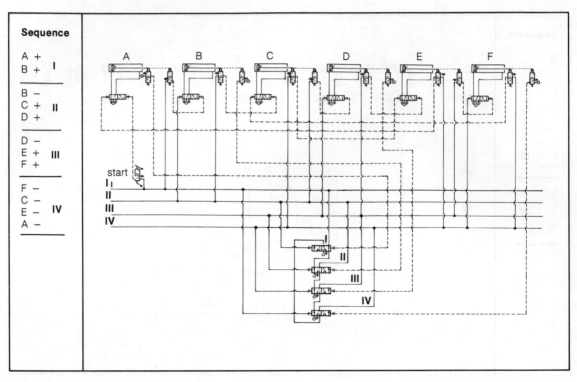

Fig. 2.24 *Connect group IV supply and signal lines*

Fig. 2.25 *Tidy up final circuit diagram*

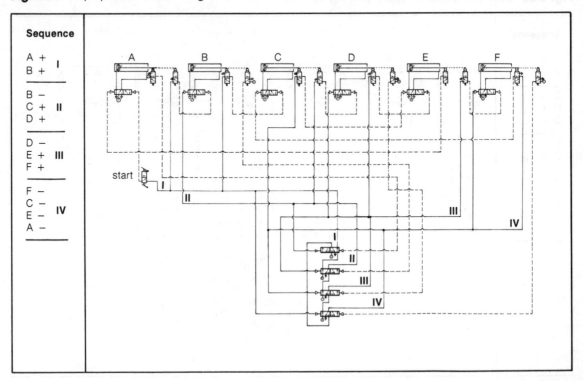

8. Now connect up those signals and supply lines required for group IV as shown in Figure 2.24.

9. The rough draft of the circuit may now be tidied up a little by erasing the unwanted portions of the supply lines drawn in the original configuration as in Figure 2.20 in step 4. Figure 2.25 shows the circuit after this tidying up has been completed.

Cascading the sequence of events for any project into groups can only be acquired through practice. At the same time, it is the habitual cascade grouping of any proposed sequence which will produce with certainty a viable structure which will form the skeleton on which the total circuit will be designed. The writing down of the sequence and grouping also forms a most useful check against any machine in which the control circuit appears to be unreliable on first acquaintance.

Adding to an existing circuit

There is often an occasion in which it is desirable to add one or two movements to an existing cycle of operations. For instance, a prototype machine, when put to work in production, demonstrates that it could be improved by adding another movement to fill a requirement which was overlooked in the original concept of the unit by the design team. In such cases, adherence to the following procedure will ensure a reliable new circuit and also save a great deal of time in the search for a viable circuit.

As an example, the following will serve to illustrate the point. The original cycle of operation was set up as:

Feed forward	$A +$	
Clamp down	$B +$	I
Drill down	$C +$	
Drill return and feed cylinder back	$C - A -$	II
Deburr forward	$D +$	
Deburr return	$D -$	
Clamp release	$B -$	III
Eject forward	$E +$	
Eject return	$E -$	I

As can be seen, this was a three-group system.

After working the machine for a few days, it was realised that the product needed a shallow slot cut in it. The mechanical arrangement of the machine was such that the most convenient way of cutting the additional slot was to move the workpiece to a second station after the deburring. Thus the additional movements comprised a transfer cylinder which held the workpiece in the second station until the slot had been cut, releasing it after the cutting so that the original ejector movement, rearranged to suit the modified mechanical arrangement, would then eject the workpiece.

When written down, the new cycle of operation took the form as shown on page 38.

The new cycle operation falls into a four-group cascade system. From every practical point of view it will be found worthwhile repiping the pilot system to meet the new group requirements. It will be *economical*, in that the circuit will require less valves; it will be *reliable*, as it will be positive with no possibility of "ghost" signals (from trapped residual exhaust); it will *ease fault finding diagnosis*. Failure to do this is often the cause of costly wasted "down time", proving the ineffectiveness of trial and error.

New cycle	Original cycle		New cycle	
Feed forward	$A +$		$A +$	
Clamp down	$B +$	I	$B +$	I
Drill down	$C +$		$C +$	
Drill return and feed cylinder back	$C - A -$		$C - A -$	
Deburr forward	$D +$	II	$D +$	II
Deburr return	$D -$		$D -$	
Clamp release	$B -$	III	$B -$	
Transfer to second station and hold			$F +$	III
Saw forward			$G +$	
Saw return			$G -$	
Transfer release			$F -$	IV
Ejector forward	$E +$	III	$E +$	
Ejector return	$E -$	I	$E -$	I

Chapter 3

Subcircuits supplementary to the cascade system

The cascade system satisfies the tests of simplicity and economy of components when the sequence is straightforward. In other words, it is ideal when, after the sequence has been written down, it can be seen that each letter appears in the cycle twice—once with a plus sign and once with a minus. If the event is a movement, it has performed a stroke in each direction. If it is something turned off or on, it will have been turned off once and on once. For such cycles the cascade method is well suited.

However, although any sequence of any other type may be cascaded to produce a circuit which will work effectively, the circuit may, at the same time, finish up as clumsy, bulky and the reverse of economical in the use of components.

Sequences which fall outside the limits of the dual appearance of each letter will prove for the most part to belong to one of three exceptions:

1. The sequence in which, while most of the letters representing events in the cycle appear twice, one letter may appear once in the first cycle then once again in the next, e.g.

First cycle
$$A + B + C + D + B - C - A - /$$
Second cycle
$$A + B + C + D - B - C - A -$$

Here D appears only once in each cycle.

2. The sequence in which all appear twice with the exception of a letter which appears four or six times or more, e.g.

$$A + B + B - B + B - C + C - A -$$

3. The sequence in which one movement must oscillate while all other events happen once each way; that is, if cylinders, they extend once and retract once, or if something is turned on once, it is also turned off once in each cycle, e.g.

$$A + B + B - B + B - \text{(oscillating as required) } C + D + D - C - A -$$

In all these cases it is usually simpler and more economical in the use of components to treat the particular letter which is out of step as a single entity within a normal cascade system, and allot to the entity a subcircuit which will cover its particular activities.

Since these several patterns are easily recognised when the sequence is written down, it becomes a matter of writing the sequence down, examining it in the normal manner, isolating the letter which is not conforming to the same pattern as the others, and determining its type of action. Then cascade the rest of the letters and allot the odd letter its particular subcircuit. Another way of describing it would be to say that the sequence is grouped in the usual cascade manner with the non-conforming letter being allotted its own kitset circuit.

The suggested method will become clearer as the following examples are traced through and examined.

Alternating signal or the letter appearing once in a cycle

The basis of the circuit dealing with this condition is shown in Figure 3.1.

Study of this circuit shows a five-port sensor which, when depressed, will give a signal which may be regarded as the action signal and is marked as such on the diagram shown. This signal is connected to the inlet port of the top five-port double pressure-operated relay valve. Passing through this valve it emerges from the outlet port to deliver an action signal which is also used for the purpose of setting the middle five-port double pressure-operated relay shown on the diagram. Setting the relay in this manner prepares this valve to pass a signal later from the sensor when it is released. When the sensor is released, the signal then given out passes through the middle five-port valve to change over the upper five-port relay valve. The upper valve is then ready to deliver the next action signal from the sensor through its second or

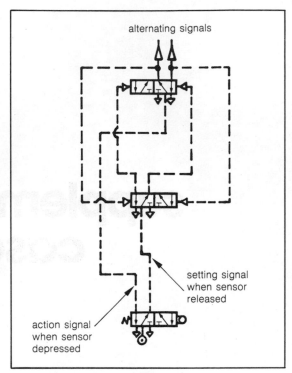

Fig. 3.1 *Subcircuit, or kitset circuit, for single movement per cycle—basic structure*

alternative outlet port. This second signal, like the first, both delivers its action signal and at the same time resets the middle relay in the diagram, preparing it for the release of the sensor, when once more a setting signal from the sensor will pass through it to reset the top five-port valve.

This basic structure may be applied to attain extensions of the idea of using alternative signals in a predetermined rotation or sequence, in many different ways. While there may be simpler ways of achieving a desirable sequence, which a combination of several of these basic structures may be able to do, the presence of other factors in the particular circumstances may render the other possible means less practicable. It is, therefore, worthwhile having such an alternative as suggested above to fall back on.

The basic structure described so far becomes a useful adjunct to the cascade system of sequential control when dealing with the question of a letter which appears once in a cycle of operation when the sequence is written down. While such a sequence may still be achieved in the conventional cascade manner, it will be found simpler to use, in slightly different form, the basic structure shown in Figure 3.1.

When applied to the movement of a cylinder, the form shown in Figure 3.1 may be recognised again in the diagram shown in Figure 3.2.

When incorporating this circuit in a cycle of several movements, the setting signal, which passes through the bottom five-port valve shown in the diagram (Fig. 3.2), must be completely exhausted before delivering the action signal, which is directed through the middle five-port valve to the correct pressure end of the cylinder's relay. If the action signal is delivered while there is still pressure in the reset signal lines, there is always the chance that the action signal may be redirected to the other pressure end of the cylinder's relay before it can be cut off. As soon as the cylinder begins to move, as may be seen in the diagram, the change of pressure condition in the cylinder changes over the reset five-port valve. Any pressure in the reset lines, naturally, will be transferred immediately to the opposite end of the relay it controls. Provided this precaution is taken in the design of the circuit, it will be found to be a thoroughly reliable circuit design.

Practical example

As a practical example of the use of this sub-circuit, or kitset circuit, in a cascaded sequential control circuit, a simple form of blow-moulding machine in a plastics factory will serve.

In this example, the cycle of operation has the following movements:

A + (mould closes)
B + (blow pipe extends into mould and extruded plastic tube now captured in the mould)
C + (knife severs the extruded plastic tube between mould and extruder)
D + (blow turned on)
delay of five seconds
D − (blow off)
B − (blow pipe retracted)
A − (mould opens)

Following the cascade rules, this sequence would be written down:

A + B + C + D + D − B − A −

Looking at the sequence as now written down, it is obvious that C has been left extended at the end of that cycle. There is a good practical reason for this. When cutting the hot, soft tube of plastic which is continuously emerging from the extruder, the knife can make only one cut at a time. To return the knife after cutting would be to drag it through the plastic tube which had been extruded after the knife made its first cut. Thus, such a knife is double-edged and passes through the plastic one way for a cut then, in the next cycle, passes through in the opposite direction to make the next cut.

To cycle until all cylinders have retracted to the original starting point and all other events are back to the starting condition, the machine would have to cycle in the following manner:

Fig. 3.2 *Subcircuit, or kitset circuit, applied to cylinder with single movement per cycle*

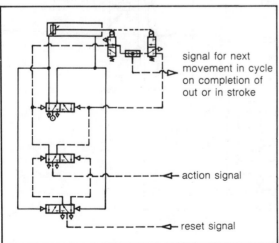

signal for next movement in cycle on completion of out or in stroke

action signal

reset signal

$A + B + C + D$ + delay 5 seconds
$D - B - A - A + B + C - D$ +
delay 5 seconds $D - B - A -$

Grouping this complete cycle in the normal cascade manner, the result would have to be:

I

$A + B + B + D$ + delay 5 seconds /

II

$D - B - A - /$

III

$A + B + C - D$ + delay 5 seconds /

IV

$D - B - A - /$

Making a four-group system of it entails directing signals from the same source to different destinations at different parts of the cycle. Altogether, the circuit becomes somewhat, and

unnecessarily, unwieldy. Study of Figure 3.3 shows how this occurs.

On the other hand, if the movement C is treated as an entity which may have its own little subcircuit, the overall circuit becomes a much more manageable affair. The grouping would be simply:

I

$A + B + C$ (subcircuit) D + delay 5 seconds /

II

$D - B - A -$

This becomes a two-group system as is seen in Figure 3.4.

Examination of the circuit in Figure 3.4 will show, in C's treatment, the subcircuit illustrated in Figure 3.2. Note that the action signal has its original source as line I and the resetting signal has its original source as line II. This, as pointed out earlier, is important as it gives

Fig. 3.3 *Sequence with cylinder C as single movement per cycle conventionally cascaded*

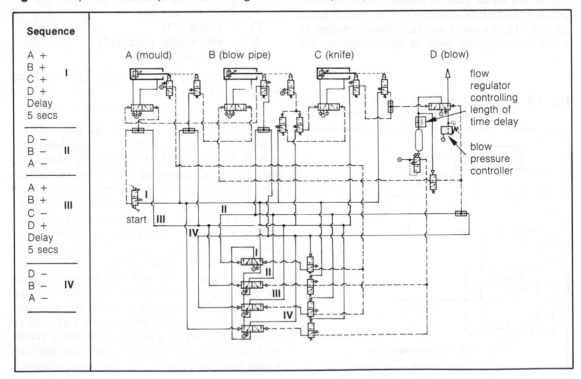

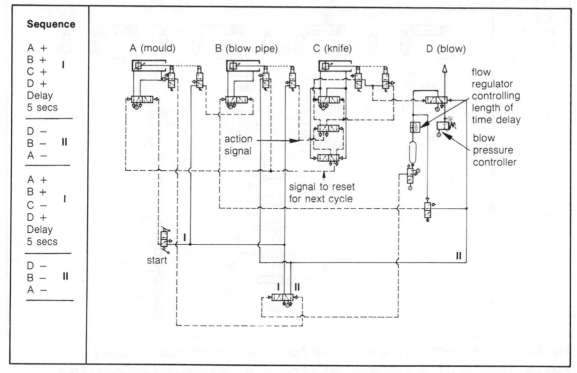

Sequence	
A + B + C + D + Delay 5 secs	I
D − B − A −	II
A + B + C − D + Delay 5 secs	I
D − B − A −	II

Fig. 3.4 *Sequence with C as single movement per cycle cascaded but also employing kitset subcircuit for C*

plenty of time to clear the lines of each from any suggestion of residual exhaust with the remote related chance of a ghost signal. Keeping the lines completely free as shown here maintains the operation as a reliable, positive, circuit with no inherent possibilities of a design malfunction.

Comparing the two circuits, Figures 3.3 and 3.4, there is a marked difference between the two in relative simplicity and number of components required. The difference becomes even greater in a cycle of operation which has more than two groups. This difference can be seen when comparing Figures 3.5 and 3.6. Figure 3.5 shows the cycle as cascaded throughout, in which event it becomes a six-group cascade system.

Figure 3.6 shows the same sequence of events grouped with the movement *B* given the subcircuit. The sequence then becomes a three-group system with a very great difference in

complexity and number of components compared with that shown in Figure 3.5. Again note the sources of the action and resetting signals in Figure 3.6.

Counting circuit

Pneumatic counting devices are available now which can be preset and will deliver a pneumatic signal when the end of the counting has been reached. A pneumatic signal will then reset it to the initial preset figure and it will start again at the beginning . This is a useful component when the number to be counted is reasonably large.

However, when the number to be counted is only two or three, a simple circuit using standard components is often useful to fall back on. The standard components required for the base

43

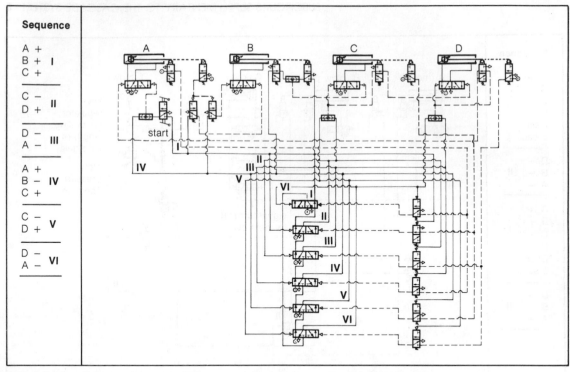

Sequence	
A +	
B +	I
C +	
C −	II
D +	
D −	III
A −	
A +	
B −	IV
C +	
C −	V
D +	
D −	VI
A −	

start

I
II
III
IV
V
VI

I
II
III
IV
V
VI

Fig. 3.5 *Sequence with B as single movement per cycle conventionally cascaded*

Fig. 3.6 *Sequence with B as single movement per cycle cascaded and employing kitset subcircuit for B*

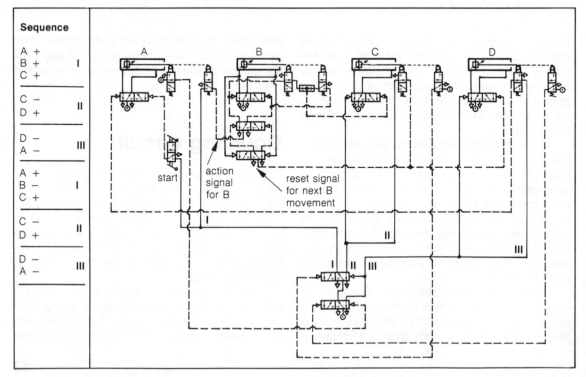

Sequence	
A +	
B +	I
C +	
C −	II
D +	
D −	III
A −	
A +	
B −	I
C +	
C −	II
D +	
D −	III
A −	

start

action
signal
for B

reset signal
for next B
movement

I
II
III

structure of this circuit are small three-port double pressure-operated valves.

Again, when this is encountered in a cycle of operations, it can be cascaded in the normal manner, but the use of the subcircuit shown in Figure 3.7 usually simplifies the system with the use of less components and a lower initial cost.

The circuit shown in Figure 3.7 is typical of the subcircuit which could be used in such a sequence as:

$$A + \underline{B + B - B + B - B + B} - C + C - A -$$

In such a sequence B has to oscillate three times before C moves. After B has completed its oscillations, it is not required to move again in the cycle of operations.

By treating B as a subcircuit, the grouping to cascade the sequence resolves into a simple two-group system:

$$\begin{array}{cc} \text{I} & \text{II} \\ A + B \text{ (subcircuit)} \ C + \ / \ C - A - \ / \end{array}$$

When drawing up the circuit, the signal from the sensor at the extremity of A's outstroke would be delivered to the inlet port of the sensor at the extremity of B's instroke as shown in Figure 3.7. After the third oscillation of B, the signal to be passed on would be applied to C's relay to initiate C's outstroke. When the pilot supply system had been changed over from line I to line II, line II would be

applied to each of the ports marked with a circle in which is the letter R. This signal would reset all the valves requiring resetting for the next cycle.

A point to note in Figure 3.7 is the use of two small differential valves to take the signal required to change over B's relay valve. The reason for these is that they will ensure that B's relay does not change over before the three-port counting valves have been changed over. If the relay moved too soon, the signal to move the counting valves could be lost before it has done its job. The differential valves selected would be such as to require a pressure signal of greater pressure than the three-port counting valves.

Predetermining number of oscillations by time delay

Where a cycle of operation requires the oscillation of a cylinder for a number of strokes, the number of which may be varied from time to time, the use of a simple delay subcircuit can provide an acceptable answer. Figure 3.8 shows the subcircuit which could be incorporated in any cascaded sequence. In this form it might be said to show the basic structure of this method.

Fig. 3.7 *Counting subcircuit for B*

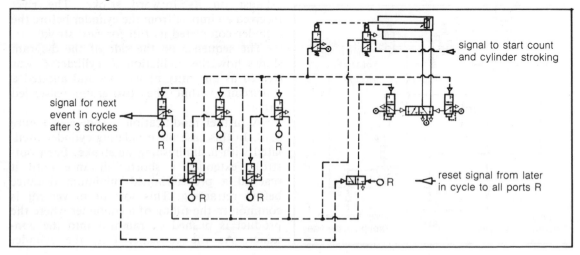

signal to start count and cylinder stroking

signal for next event in cycle after 3 strokes

R   R   R

R   R

reset signal from later in cycle to all ports R

O R

Operation

The mode of oscillation in Figure 3.8 employs a sensor at each end of the cylinder stroke, signals from which reverse the cylinder movement by changing over the relay valve. The sensor sensing the instroke extremity sends its signal to the inlet port of a small, five-port, low-pressure sensing valve—usually a small diaphragm-operated air or spring-return valve. So long as this valve is held in a depressed state with pressure on its pressure-sensing port, the signal from the sensor passes through to the relay valve controlling the cylinder, causing it to change over and extend the cylinder. When all pressure has been exhausted from the pressure-sensing port of this diaphragm-operated valve, it changes over and redirects the signal from the sensor. This redirected signal is used then to initiate the next movement in the main cycle of operations.

The initiating signal for the oscillations changes over a small three-port valve which, until then, has pressurised the reservoir and pressure-sensing head of the diaphragm-operated valve. Pressure from the reservoir and pressure-sensing head exhausts through the three-port valve at a rate controlled by the flow regulator. This method is the commonly used "bleed off" time-delay circuit.

Fig. 3.8 *Oscillating subcircuit on time delay*

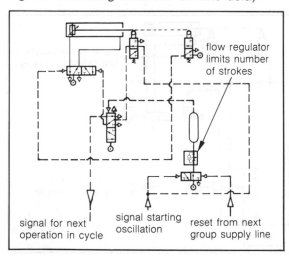

signal for next operation in cycle

signal starting oscillation

reset from next group supply line

flow regulator limits number of strokes

The three-port double pressure-operated valve used to start the timing is reset by the next supply line in the cascaded pilot supply system. For example, if the start signal of the oscillation has as its source line I, the reset signal will come directly from line II.

Modifications of the basic structure, oscillations by time delay

Modifications of the method shown in Figure 3.8 may be used to advantage for practical purposes as the following practical example illustrates.

A machine was required which would rivet light components. Because the structure of the component was not strong enough to withstand the normal heavy blow required to secure the rivet, it was decided to deliver a series of light impacts on the rivet which would eventually flatten the rivet to the desired shape. The machine was required to handle a number of different sizes of rivets so that the number of impacts, or blows, delivered was to be varied at will, within certain limits.

Figure 3.9 shows how the subcircuit was applied to this problem. Note that the sensor sensing the extremity of the rivetting cylinder outstroke is a pressure-sensitive diaphragm-operated valve sensing the decay of the cylinder exhaust on its forward stroke. The rivet received an impact from the cylinder before the cylinder completed its full forward stroke.

The sequence on the side of the diagram shows how the oscillation of cylinder *C* was treated as an entity of its own and allotted a subcircuit within the two-group cascaded circuit.

A further modification is shown in Figure 3.10. In this case the oscillating cylinder oscillated with a diminishing outstroke. Each outstroke extended a shorter distance until it reached a predetermined minimum distance before retracting. This sort of movement is common to the filling of a container where the product is pushed or rammed into the container. As the container fills, the cylinder

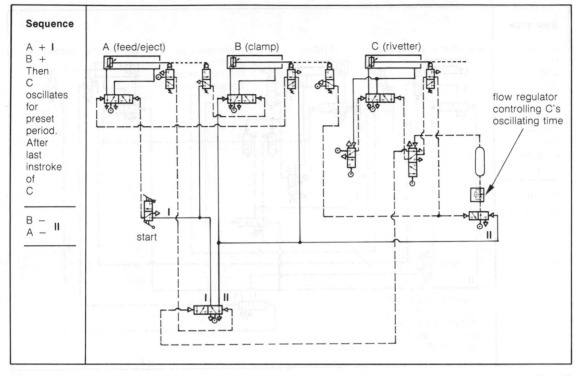

Fig. 3.9 *Practical example of oscillating subcircuit*

Fig. 3.10 *Oscillating cylinder with outstroke diminishing to predetermined point*

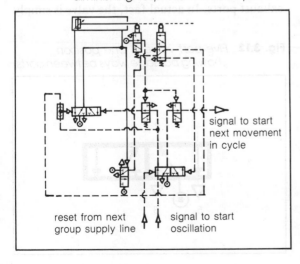

pushing the product in finds its stroke progressively shortened.

Study of the circuit in Figure 3.10 shows how the sensing of the forward movement is carried out by a pressure-sensitive diaphragm-operated valve so that the cylinder drives forward as far as it can then returns on the decay of the forward exhaust of the cylinder. The predetermined position for the final minimum outstroke—completion of the filling operation—is marked by the five-port roller-operated pilot valve. When the signal to return emanating from the diaphragm-operated valve coincides with the position of the cylinder, in that it has stopped at the point where the five-port roller-operated valve is depressed, the initiating signal, which was sustained and passed through the small five-port double pressure-operated valve, is cut off and a new signal is directed to initiate the next movement in the overall cycle.

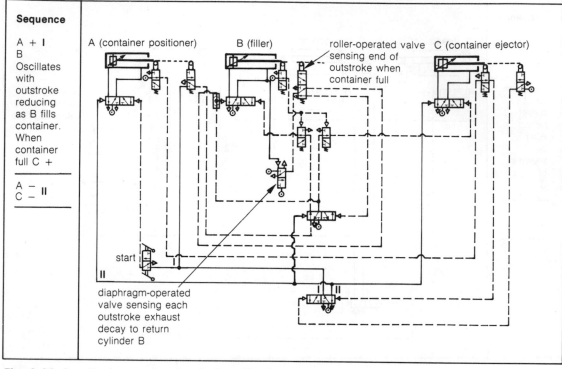

Sequence

A + I
B
Oscillates
with
outstroke
reducing
as B fills
container.
When
container
full C +

A – II
C –

A (container positioner)   B (filler)   roller-operated valve sensing end of outstroke when container full   C (container ejector)

start

diaphragm-operated valve sensing each outstroke exhaust decay to return cylinder B

Fig. 3.11 *Practical example of cylinder with diminishing oscillating outstroke*

Figure 3.11 shows the subcircuit for this incorporated in a typical practical example of this sort of operation. Note how the oscillation of cylinder *B*, the filler cylinder, has been treated as an entity on its own within the two-group cascaded system.

Other useful kitset subcircuits

The subcircuit about to be described is typical of the use of standard pneumatic valves where the valves are regarded simply as directional flow valves rather than specific types for specific functions. For example, the novice in the world of pneumatics can find, unwittingly, that the five-port valve has come to be looked upon as a valve with one inlet port, two outlet ports and two exhaust ports. Let Figure 3.12 act as a working example.

In Figure 3.12, the flow of air is shown as always in through the inlet port, 1, and out through either of the two outlet ports, 2 and 4. The other two ports, 3 and 5, are shown as exhaust ports. In actual fact, the valve is simply

Fig. 3.12 *Five-port valve in one position showing passage ways between ports*

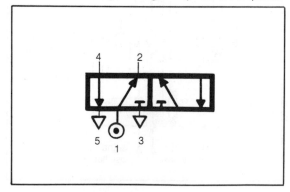

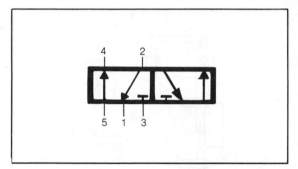

Fig. 3.13 *Five-port valve in alternative position showing passage ways between ports*

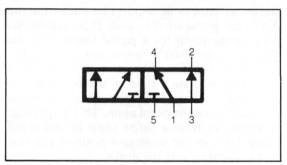

Fig. 3.14 *Five-port valve as directional flow valve, drawing individual air supply for each outlet port from two separate pressure sources*

a device which in one position offers a passageway through from port 1 to port 2 in either direction, 1 → 2 or 2 → 1, and also while in this position there is a second passageway between ports 4 and 5 in either direction, 4 → 5 or 5 → 4. Port 3 is closed off altogether with the valve in this position.

When the valve is changed over to its second position, the passageways through will then be between ports 1 and 4 and ports 2 and 3, while port 5 will be closed off altogether.

Using the valves as directional flow valves, the arrows showing the flow will vary as the occasion demands. The low pressure return of a cylinder with no load on its return stroke, discussed in *Practical Pneumatics*, provided an illustration of the use of a five-port valve departing from the concept shown in Figure 3.12.

Once this broader concept of pneumatic valves has been grasped, the logic functions of these valves in circuit design are readily understood and their versatility becomes a useful tool in the hands of the imaginative designer.

Typical of this concept is the basic structural circuit shown in Figure 3.15.

Examination of this circuit will show that whichever of the two valves is changed over, the end result will be a reversal of the flow of air to the cylinder itself. From a practical point of view, this can have many uses. It may be applied directly to an air cylinder or to the dash-pots of a hydro-pneumatically operated cylinder.

Practical example

The circuit shown in Figure 3.15 has proved useful in many sawmills in the control of what are called "docking saws". A simple form of such a saw is operated in the first instance by a foot-pedal operated, air/spring-return five-port valve. The saw is required to extend only as far as a predetermined point in its outstroke. Reaching this point it is then required to retract immediately. If, however, the operator decides the saw should retract before it reaches the point of return which has been set, the saw will

Fig. 3.15 *Kitset basic structural circuit for alternating relay valves*

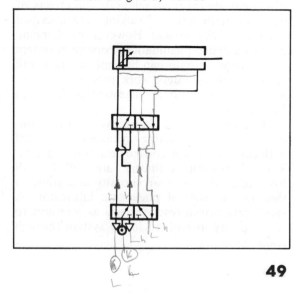

49

retract as soon as he removes his foot from the operating pedal of the valve. If he retains his foot on the pedal for a period longer than the time taken to extend the saw and retract fully, the saw will return at the point set then remain retracted until such time as the operator has removed his foot completely from the pedal then depressed the pedal again. This latter point is important from a safety angle in that if the pedal jams in the depressed position the saw will make only one outstroke.

Figure 3.16 shows the circuit for this operation. Note that the upper five-port valve is reset when the foot pedal is released and the system is then ready for the next cycle. Without this resetting signal there can be no further movement of the cylinder.

It should also be noted that both relay valves are of the same port size so that the speed of the cylinder is not affected by one valve being undersized.

Control of two cylinders moving precisely together

As is generally known, it is not possible to make two air cylinders work precisely together without some form of mechanical arrangement which will tie all the movements of the cylinders together so that they work as one. The steps which can be taken to minimise the effects of what is usually termed "walking" are described in *Practical Pneumatics*. However, by changing over to a hydro-pneumatically operated system an equalising circuit can be applied which will ensure both cylinders work precisely as one. The circuit for this arrangement is shown in Figure 3.17.

In Figure 3.17 are shown two tandem cylinders, i.e. double-acting cylinders connected with common piston rods to dash-pot cylinders. These dash-pot cylinders are filled with hydraulic oil of similar viscosity and grade to that in the normal pneumatic lubricator. A small pressurised reservoir acts as a reserve to take up any loss of oil in the system through leaks.

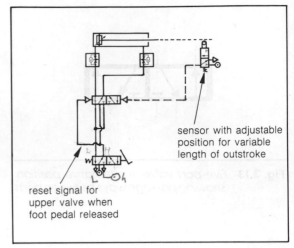

sensor with adjustable position for variable length of outstroke

reset signal for upper valve when foot pedal released

Fig. 3.16 *Practical example of applied kitset for alternating relays*

Operation

The lever-operated five-port valve supplies air to the air cylinders, driving them either forward or in reverse.

The moving air cylinders drive, in turn, the dash-pot cylinders. With the oil line connections as they are on the diagram, the displacement oil from one side of the piston in one dash-pot cylinder is driven to the other dash-pot cylinder's opposite end. In this way, it is not possible for one cylinder to move without the other receiving the displacement oil from the first, simultaneously. To receive this displacement oil simultaneously, the displacement rate of both cylinders must be the same.

Discrepancies can occur when the cylinder seals become worn. When this occurs the circuit arrangement provides for immediate resynchronisation at the end of each stroke. If one or other of the sensors marked *A* is not depressed at the end of the movement, either valve *B* or *C* will pressurise the signal port of valve *D*, causing it to open and bring both cylinders back into phase.

The flow regulators act as non-return valves to prevent a free flow of displacement oil into the reservoir.

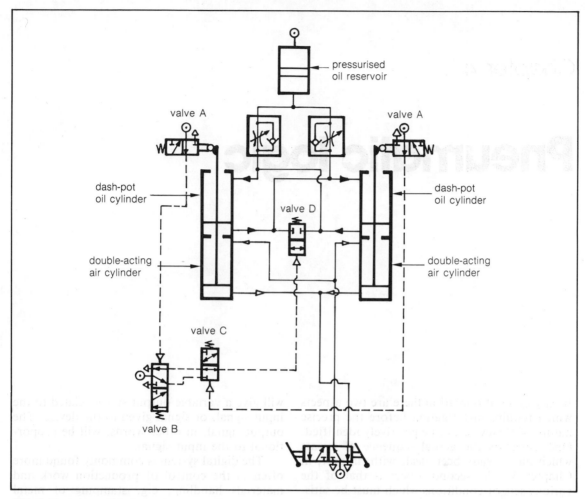

Fig. 3.17 *Kitset subcircuit for near simultaneous movement of two hydropneumatically operated cylinders*

Chapter 4

Pneumatic logic

In any cycle of operation there are two aspects which require investigation before the precise nature of the cycle can be positively identified. One aspect is the actual sequence of events which has already been dealt with, in part, in Chapter 3. The second aspect is that of the various sets of conditions which must be satisfied before a further step in the sequence of events may be taken. It is this second aspect which is generally implied when reference is made to pneumatic logic.

A further implication of the term "pneumatic logic" in this context is that the process control system is of a digital rather than an analogue nature.

A digital system is one employing digital components or devices. These devices will have two positive states. Which of the two states it may adopt will depend on the presence or absence of a control signal or, again, on the last control signal it received.

An analogue system is one employing analogue components in which the devices used will give a variable output signal related to the input signal, or signal given to the device. The output signal, in other words, will be proportional to the input signal.

The digital system is commonly found more often in the control of production work and materials handling, e.g. stamping of metal components, cutting and trimming, pushing workpieces from one position to another. The analogue system is more commonly found in continuous flow processes such as the pulp and paper process or the manufacture of fibrous plaster board. Digital is more of an on/off system, while the analogue is proportional.

Digital devices may be classified under one or other of two main categories, bistable or monostable.

A bistable pneumatic device is one which will remain stable in either of its two alternative states even though the signal which caused it to take up a particular state is removed. It will remain so until a further signal is delivered to change it over to its alternative state.

Sometimes such a device is referred to as a *memory* device. Such devices are typified by a double pressure-operated spool valve.

A monostable pneumatic device is one which will revert to its original state as soon as the signal which changed it over to its alternative state is removed. Such a monostable pneumatic device is typified by air-operated spring-return valves or by differential valves with air return.

Both types of device may be further classified under two general categories—active and passive. *An active pneumatic device* is one which has a permanent air supply to its inlet or input port. *A passive pneumatic device* is one which has intermittent pressure supplies or signals applied at appropriate times to its inlet or input port.

Both of these latter terms, active and passive, may be applied to one and the same component, the term applied being dependent on the manner in which the component or device is used in the circuit. This will be seen more clearly as the various logic structures are described in this chapter.

Methods of identifying qualifying sets of conditions

In considering pneumatic logic structures, it should not be forgotten that in practice there will be many transient conditions in circuits while valves are being switched from their "off" to "on" positions and vice versa. There is always a period, no matter how short, in which pressure for a signal is building up or a signal no longer required disposes of or "exhausts" its working pressure.

In the initial stages of a circuit design, however, the matter of transients and response times, together with the actual physical nature and final selection of components, will be left until later. The first important step is to identify positively, then write down, the details of each of the sets of conditions which prevail at each stage of the cycle of operation.

In a straightforward sequence in which each movement must be completed before the next movement starts, the set of conditions is simple, requires no further amplification and can be adequately catered for by the cascade system of sequential control. When, however, additional conditions are introduced which will qualify the operation in some way, it becomes necessary to resort to some form of logic structure which will couple these conditions with the main sequential structure already drawn up. If it is necessary to ensure that the raw material is there and a guard in place before an operator presses a start button to start an operation, two conditions, over and above the sequence of operation, have been introduced. They qualify the effective start of the operation. The material must be there and the guard must be in place and the start button must be pressed before the first movement will take place. This set of conditions is generally described as an AND logic function. The appropriate valving would be selected for this condition and added to the skeleton circuit design of the main sequence.

The systematic analysis of the basic form or structure of sets of qualifying conditions is generally attributed to George Boole, (1815–64), an English mathematician. He pointed out the analogy between algebraic symbols and those representing logical forms and syllogisms which existed in philosophical reasoning and thought.

Among the many papers he published, two in particular contributed to the establishment of fundamental mathematical laws of logic—*Mathematical Analysis of Logic*, 1847, and *An Investigation into the Laws of Thought, on Which Are Founded the Mathematical Theories of Logic and Probabilities*, 1854.

From these and subsequent investigations a useful concept for practical purposes has evolved which provides clear identification of the common combinations of qualifying conditions. Each combination has been given a descriptive name and its individual logic circuit structure.

Eleven such basic logic forms are listed in the pages which follow. To enable those more familiar with other forms of control to relate to the practical pneumatic approach, the logic forms are listed under their normal logic terminological titles, together with their appropriate truth tables, symbolic and Boolean algebraic equations, electrical circuit equivalents and, of course, their pneumatic circuit alternatives.

It will be noted that, in most cases, two alternative pneumatic circuits are shown. The final selection of one or the other depends on the physical nature and layout of the particular project with which the circuit is involved. Suggested guidelines will be discussed later.

The truth tables are the conventional means of showing the different output conditions which will result from all possible input conditions. The figure **1** denotes the existence of the signal and **0** denotes its non-existence. **S** is shown as the resulting output signal. **X** and **Y** denote input signals.

The NOT function (negation)

Graphic symbol

Algebraic equation   X =

Truth table

X	S
1	0
0	1

If X exists, then S does NOT exist. If S exists, then X does NOT exist.

Logical circuits

pneumatic   electrical

The OR function (disjunction)

Graphic symbol

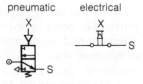

Algebraic equation   X + Y = S

Truth table

X	Y	S
0	0	0
1	0	1
0	1	1
1	1	1

If X exists or Y exists, or both X and Y exist, S exists. If neither X nor Y exists, then S does not exist.

Logical circuits   pneumatic   electrical

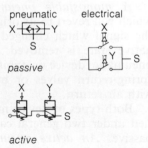

passive

active

The AND function (conjunction)

Graphic symbol

Algebraic equation   X • Y = S

Truth table

X	Y	S
0	0	0
1	0	0
0	1	0
1	1	1

If both X and Y exist, then S exists. If either X or Y does not exist, then S does not exist. If both X and Y do not exist, then S does not exist.

Logical circuits   pneumatic   electrical

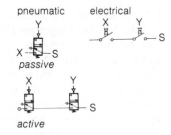

passive

active

The NOR function

Graphic symbol

Algebraic equation $\overline{X + Y} = S$
                            $\overline{X} \cdot \overline{Y} = S$

Truth table

X	Y	S
0	0	1
1	0	0
0	1	0
1	1	0

If neither X nor Y exists, then S does exist. If either X exists while Y does not exist, or Y exists while X does not exist, then S does not exist. If both X and Y exist, then S does not exist.

Logical circuits   pneumatic   electrical

active

active

The MEMORY function

Graphic symbol

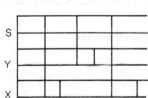

Function diagram

S

Reset signal   Y

Set signal   X

A memory unit is a bistable device which provides a convenient means of storing logical conditions. Once it has changed over to the appropriate state on receipt of a signal to do so, it will retain that state even though the signal is removed. It will remain so until a signal is received to change it over to its alternative state. While the diagram below shows a double pressure-operated five-port valve, the double pressure-operated three-port valve may also be regarded as a memory device.

Logical circuits   pneumatic   electrical

The IDENTITY function

Graphic symbol

X ———▷ S

Algebraic equation   X = S

Truth table

X	S
0	0
1	1

If X exists, then S exists. If X does not exist, then S does not exist.

Logical circuits   pneumatic   electrical

The NAND function

Graphic symbol

Algebraic equation $\overline{X \cdot Y} = S$
$\overline{X} \cdot \overline{Y} = S$

Truth table

X	Y	S
0	0	1
1	0	1
0	1	1
1	1	0

If both X and Y do not exist, then S exists. If X exists and Y does not exist, then S exists. If Y exists and X does not exist, then S exists. If X and Y both exist, then S does not exist.

Logical circuits   pneumatic   electrical

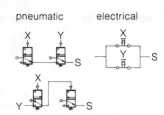

If both X and Y do not exist, then S exists. If X exists and Y does not exist, then S exists. If X does not exist and Y exists, then S does not exist. If both X and Y exist, then S exists.

The INHIBITION function

Graphic symbol

Algebraic equation $X \cdot \overline{Y} = S$

Truth table

X	Y	S
0	0	0
1	0	1
0	1	0
1	1	0

If X exists and Y does not exist, then S exists. If both X and Y exist, then S does not exist. If Y exists and X does not exist, then S does not exist. If both X and Y do not exist, then S does not exist.

Logical circuits   pneumatic   electrical

passive

active

The IMPLICATION function

Graphic symbol

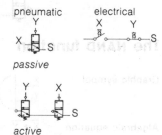

Algebraic equation $X + \overline{Y} = S$

Truth table

X	Y	S
0	0	1
1	0	1
0	1	0
1	1	1

If both X and Y do not exist, then S exists. If X exists and Y does not exist, then S exists. If X does not exist and Y exists, then S does not exist. If both X and Y exist, then S exists.

Logical circuits   pneumatic   electrical

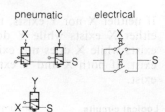

The THRESHOLD function

Graphic symbol

Algebraic equation $\overline{X} \cdot \overline{Y} + X \cdot Y = S$

Truth table

X	Y	S
0	0	1
1	0	0
0	1	0
1	1	0

If both X and Y do not exist, then S exists. If X and Y both exist, then S exists. If X exists and Y does not exist, then S does not exist. If Y exists and X does not exist, then S does not exist.

Logical circuits   pneumatic   electrical

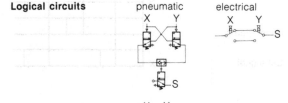

The EXCLUSIVE OR function (antivalence)

Graphic symbol

Algebraic equation $\overline{X} \cdot Y + X \cdot \overline{Y} = S$

Truth table

X	Y	S
0	0	0
1	0	1
0	1	1
1	1	0

If both X and Y exist, then S does not exist. If both X and Y do not exist, then S does not exist. If X exists and Y does not exist, then S exists. If Y exists and X does not exist, then S exists.

Logical circuits

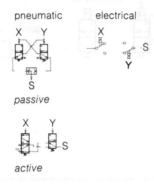

pneumatic   electrical

passive

active

Expansion of the basic logic form

All of the examples given are in their minimal form. They can be expanded as required to include many other sources of signals. For example, before a hopper gate, about to release aggregate on to a conveyor belt for transfer to a concrete mixer in a batching plant, may be allowed to open, it would be necessary to ensure that the following conditions were already fulfilled:

1. the conveyor electric motor was running.
2. the mixer gate was open to receive the aggregate.
3. the mixer discharge gate was closed.
4. the aggregate hopper was full of aggregate.
5. the mixer motor was running.

Since all conditions would be sending out a positive signal indicating compliance with requirements, this would clearly be an AND logic circuit. The circuit would then be drawn up along the lines of that shown in Figure 4.11.

Before looking at Figure 4.11, consideration should be given to the question raised earlier as to which of the examples given, active or passive, should be used. In this case, obviously some of the signals would be coming from quite some distance from the control panel. In the cases of the two motors, the signals would in all probability be electrical. Since there are five conditions to be met, apart from the start signal which may come from another source again, the question of strength of signal which is finally delivered to the valve which must action the hopper gate prime mover must be considered. Thus the active alternative could be used to advantage in fulfilling the need for a good strong signal, combined with the ever present need for economy in the use of valves.

Figure 4.11 can be seen to have taken these factors into account.

Another practical example, which will serve to illustrate the point raised in respect to expanding the basic structure, is that of a high-speed planing machine in a timber mill.

In this case the timber is driven through the planing knives by a series of pressure-drive rollers. With the timber passing through at high speed, the need for an emergency release of the pressure-drive rollers had to be met in the event of a break-up in the timber. Without such an immediate release of the drive rollers, damage to the machine could be expensive. To meet this, five pneumatic sensors were installed at intervals along the length of the machine to sense any unscheduled change of position of the timber passing through. A signal from any one of these five sensors changed over the relay valve controlling the air cylinders exerting pressure on the drive rollers. Figure 4.12 shows how the shuttle or tee valves were arranged to convey these signals to the action relay. Clearly here was an OR logic situation.

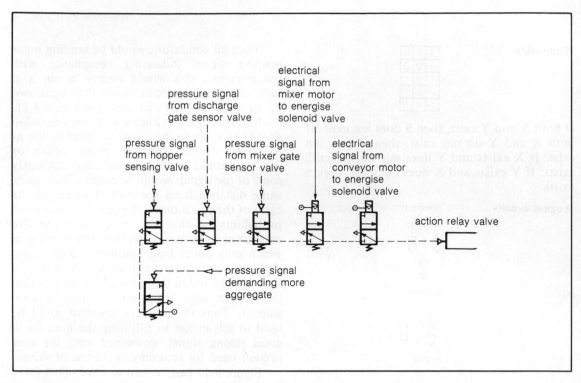

Fig. 4.1 *Circuit for a hopper gate*

Fig. 4.2 *Shuttle or tee valves arranged in* OR *logic situation*

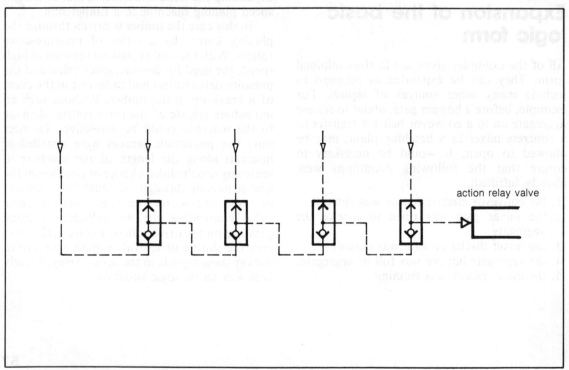

Chapter 5

Alternatives to the cascade system

Disposal of the unwanted signal has been dealt with in a number of ways apart from the cascade method. In more recent times, equipment manufacturers have striven to introduce logic systems which are claimed to be different. To the uninitiated such systems may appear to be so. Closer investigation generally shows an underlying common denominator in most of them. The surface differences are mainly in the mathematical approach each has employed in an attempt to produce a workable system.

Concurrently with the search for effective circuit design methods, development of components has taken place along similar lines in many different countries. Components for logic control have been considerably reduced in physical size. With miniaturisation has also come further development of glandless or sealless valves. The use of diaphragms to achieve fast response and sensitivity to low-pressure signals has become widespread. The overall picture in modern logic components is one of response times measured in terms of one to five milliseconds or so, repetitive consistency of such a high order as to contain barely measurable variations in response times, sensitivity to pressure changes of three or four kPa and to movement of 0.025 mm to 0.05 mm.

Some earlier methods of disposing of unwanted signals, because of their inherent limitations, have fallen into disuse to a large extent. However, they have been included in the following review of the various alternatives as much from historical interest as from any practical value they may have. Again, while the types discussed do not cover by any means all the alternatives available, they form a fairly comprehensive representation of the methods employed by most of the methods to date. The objective is not to present any system in fine detail so much as to describe the underlying general principles on which the various types are based. With an understanding of these, the various methods offered will be readily understood when encountered in their various superficial guises.

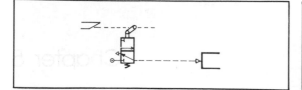

Fig. 5.1 *ISO symbol for one-way roller trip sensor*

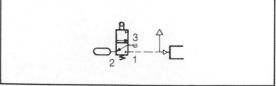

Fig. 5.2 *Precharged reservoir sensor*

One-way roller trip sensor

This type of sensor is operated by a spring-loaded device which allows a cam to depress its lever arm and the valve mechanism in one direction only. When the cam moves across the device on its return movement, the device gives way to the cam without depressing the valve mechanism. After the cam has passed over, the spring-loaded device returns to normal. Figure 5.1 shows the CETOP symbolic form of such a device.

Although such a sensor will produce a signal which is of limited duration, it must be used with discrimination. Its limitations need to be fully recognised and matched against the function required by the particular operation before it is included in any project. These limitations are briefly:

1. The duration of the signal is governed by two factors:
   (a) the length of the cam;
   (b) the speed of the moving cam.
2. The signal produced from such a sensor can be produced only while there is movement. If the project demands that a signal should be given at the end of a movement, and also that the signal be of limited duration so that it does not present problems with its presence later in the cycle, this cannot be performed by such a device.

Precharged reservoir method

This method, of rather doubtful value, is shown in Figure 5.2. In this case the signal is of limited duration. A sensor, as shown in the diagram, of the three-port, roller-operated spring-return type is used. Port 3 is connected to the constant air supply; port 1 is connected to the pressure end of the valve receiving the signal; port 2 is connected to a small reservoir.

In the released position, port 3 is open through the valve to port 2. While in this position the reservoir is charged with its supply of pressurised air. When the roller mechanism is depressed, port 3 closes and port 2 is connected through the valve to port 1. The compressed air in the reservoir expands into the pipe connecting port 1 to the pressure end of the valve receiving the signal. This pressure in the pipe must be exhausted before the valve receiving the signal may be changed over to its original position by a signal on its opposite pressure end. Accordingly, the pressure is exhausted to atmosphere through a small bleed hole as seen on the diagram. As a rule, where such a method as this has been used, the hole has been bored either in the connecting pipe or in the fitting holding the pipe to the pressure end of the valve taking the signal. As can be imagined, there are several factors with direct influence on the success or failure of this method.

The reservoir, generally a standard size, must be matched with the length of the signal pipe to ensure that, when expanded, the air released from the reservoir has sufficient pressure to change over the valve receiving the signal. This means checking and calculating:

1. the minimum pilot pressure required by the particular valve receiving the signal. (Minimum pilot pressures for different valves can vary within quite wide limits; they can be as high as 350 kPa.)
2. the volume of the pipe selected (area of bore times length) to convey the signal.

3. time required to move over the operating mechanism of the receiving valve.
4. additional volume created during the changeover of the receiving valve (displacement volume of its spool or what-have-you).
5. time which may be allowed for the signal pressure to exhaust completely through the bleed hole.
6. the size of the bleed hole which should be sufficient to allow exhausting of the signal without reducing the pressure of the signal below its critical level before the receiving valve has changed over. In calculating the size of this hole it must not be forgotten that to exhaust down to atmospheric pressure again there will be a period of sonic flow then a further variable subsonic flow over a second period of time.

Taking all these factors into consideration, it can be seen that a precise answer and a reliable signal will be difficult to achieve under average practical conditions. At best, this method could be regarded as an emergency makeshift method to keep something going until the means of applying a positive, reliable method are available.

Pulse valve, fixed duration

Several manufacturers offer a component which can deliver a signal of limited duration. At the end of its period, the signal is cut off, and the pressure, delivered as a signal, is exhausted to atmosphere back through the device.

Figure 5.3 shows the general arrangement which carries out the function. In the components offered as a single device the arrangement is integrated inside the component.

As may be seen from Figure 5.3, air from the pressure applied to the inlet port both passes through the device to deliver the required signal and also is used to cause the mechanism in the device to change over. Changing over of

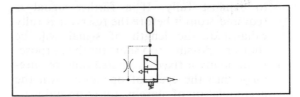

Fig. 5.3 *Pulse valve sensor arrangement*

the mechanism cuts off the signal and exhausts pressure downstream from the device. To maintain the signal long enough before cutting it off, the pressure applied to the mechanism is bled through a restrictor into a small reservoir, thence to the operating pressure end of the mechanism. Pressure builds up slowly to the changeover point. In fact, the method is identical to that known as the bleed-on time delay.

Figure 5.4 shows a single pulse arrangement in which the length of the pulse may be varied. The restrictor in this case is adjustable.

In practice, there are several questions which should be asked before this method of avoiding the problem of opposing signals is used.

1. Is it safe to assume that the length of time in which the signal is on is sufficient to allow for the build-up of pressure in the pipelines through which the signal must travel to its destination?
2. Has the valve accepting the signal a consistent response time which will allow for its changeover before the signal is cut off and exhausted?
3. How long is the time required by the device to prepare for its next operation? This time will depend on how long the reservoir takes

Fig. 5.4 *Variable pulse length valve arrangement*

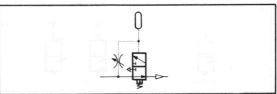

to exhaust fully. If a further signal is required from it before the reservoir is fully exhausted, the length of signal will be shorter—possibly too short for the purpose.

4. In applying a fixed estimated time requirement into the cycle of operation, will the accumulation of margins for error and variables be detrimental to the production potential of the machine using this device in its control system? As pointed out earlier, such an accumulation of a number of margins, each of only a few milliseconds, can result in serious reductions in production.

Another characteristic of this method which warrants some consideration before making a final decision is in fact that it turns itself off. In the event of an unscheduled stoppage, diagnosis of the cause of the stoppage is not assisted when there is no means of telling whether or not a signal was actually delivered.

Use of AND and NOT components

Both AND and NOT components are often advocated as means of overcoming the problem of the unwanted signal or of delivering a signal only when a particular combination of conditions exists.

Figures 5.5 and 5.6 show these two types of approach. The AND uses the method of limiting the delivery of the signal to specific conditions and the NOT eliminates the unwanted signal causing a problem of opposing signals on the same relay valve.

There is no question as to the effective performance of each of these when considered

in isolation from the context of a whole control circuit for a cycle of operation. On the other hand, indiscriminate use of this method of achieving a sequence of operation can cause the final complete circuit to become unnecessarily involved. Such an approach will often result in the use of more components than will be required by a cohesive cascade approach. Diagnosis of unscheduled stoppages becomes more difficult when an unnecessarily large number of components is used with no clear pattern of logical thought behind their application.

As with some methods discussed earlier, unnecessary components add those unproductive milliseconds which detract from the potential production of the machine controlled by such a system.

Rotating mechanical cams

The mechanical arrangement shown in Figure 5.7 forms the basis of many different devices offered by manufacturers as a means towards sequential operation.

The camshaft is driven by a motor of some sort—electrical or pneumatic. Usually the motor drives the camshaft via a reduction gear assembly. The speed of rotation is set to relate to the time taken to perform one complete cycle of operation. As the shaft rotates, the cams, which have been adjusted to their appropriate intervals, depress microvalves. The same method is employed using electrical motors and microswitches where the control system is electrical rather than pneumatic. Cam lengths are

Fig. 5.5 AND component arrangement

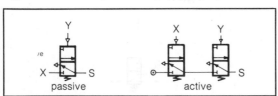

Fig. 5.6 NOT component arrangement

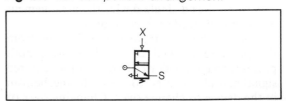

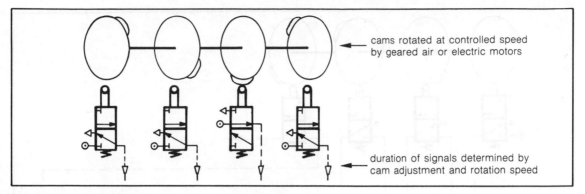

cams rotated at controlled speed by geared air or electric motors

duration of signals determined by cam adjustment and rotation speed

Fig. 5.7 *Rotating mechanical cam arrangement*

usually adjustable so that they can depress the microvalves long enough for them to give out usable signals. Naturally cam lengths must be related to the likely speed of rotation. Speed of rotation may be fixed or variable.

The degree of complexity such units achieve varies from a single rotating shaft with ten cams and microvalves or less to three or four interconnected shafts, each driven at a speed which is a multiple of the main drive shaft. Associated cams and microvalves on such types can amount to a considerable number.

In this form, such units depend on an initial estimate of the length of time each section of the process may take. The estimate must include a reasonable margin for unforeseen variables. The sum of these estimates then provides the time allowance for the whole cycle. This total time provides the setting for the speed of rotation of the camshaft. As pointed out earlier, the time taken up in waiting for the signal for the next section of the process to start can be an appreciable percentage of potential production time when compared with a feedback control system which allows each section to start as soon as the preceding section has been completed.

Another aspect of this type of control is the fact that if a particular section of the process runs into trouble, the next section will receive its initiating signal regardless of whether the previous section is complete or not. This can result in considerable mechanical damage in some types of machinery controlled in this way.

Because of this last aspect, there are modified versions of this concept available, employing some form of interlocking.

Interlocked rotating mechanical cams

Figure 5.8 shows a commonly used method of protecting a machine from the possible mechanical damage which could be caused by the cycle continuing in spite of a malfunction.

Looking at Figure 5.8 from left to right, it can be seen that a signal is given to the left-hand cylinder to extend. When it has completed its outstroke, a sensor at the extremity of its outstroke is depressed. The output from this sensor

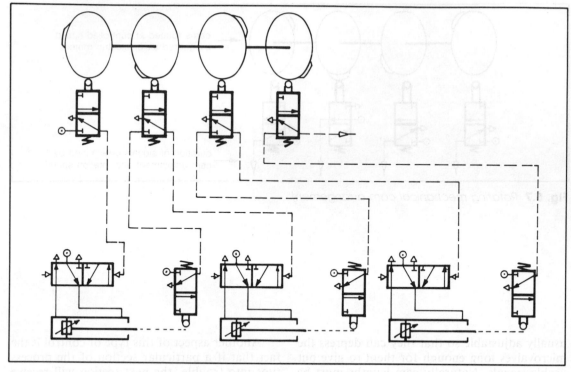

Fig. 5.8 *Interlocked rotating mechanical cam arrangement*

is applied to the inlet port of the sensor on the rotating cam device which is waiting to be depressed by a cam and will provide the signal for the second cylinder to extend. This ensures that the second cylinder cannot move forward until the first cylinder has extended. If the first cylinder is late in arriving at the end of its outstroke, there will be no air supply for the signal initiating the second movement. From this it can readily be seen how an allowance must be made for all possible contingencies in setting the speed of the rotating cams. It must be set so that there is time for air always to be waiting at the inlet port of each sensor before its respective cam approaches.

Although such a method may correctly claim to incorporate "feedback", the feedback is not used to full advantage in that it is not used to initiate the next movement—eliminating unnecessary delay.

Feedback, pulse-propelled mechanical cams

Figure 5.9 shows a further extension of the rotating mechanical cam method.

In this case, the camshaft is rotated with a ratchet and pawl mechanism. The mechanism is propelled by a small single-acting air cylinder. The cylinder is driven forward by a limited duration pulse signal. This signal is derived from a sensor which indicates that the preceding movement, or section of the process in the cycle, has been completed.

Such a device allows for quick and easy changing of a sequence. Adjustment of the cams is all that is required to change the sequence, or to reprogram the cycle of operation. This attribute can be very useful when

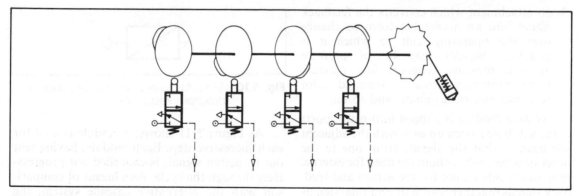

Fig. 5.9 *Feed-back, pulse propelled, rotating mechanical cam arrangement*

developing a prototype machine. As many well know, the original concept of the machine often undergoes modifications as the development proceeds. Sometimes an additional movement is found to be required, or another may be required to operate later in the cycle. In such circumstances, the use of a device of this nature will effect savings in both labour and materials when compared with the repiping of a modified circuit.

Later, when the prototype has been proved and the final developed machines are ready to be produced as standard units, the reaction time and repetitive consistency of such a unit will need to be compared in terms of cost and productivity with any alternatives.

Modular program sequencers

An entirely different approach to the problem of attaining an effective sequential control without the possibility of signals opposing each other on either end of a double pressure-operated relay valve is provided by the modular program sequencer.

Each module has an output port which provides a signal for the next step in the cycle of operation and an input port which receives the feedback signal to indicate that the required action has taken place.

On receipt of the feedback signal, the module supplies a signal to the module initiating the next step. At the same time the signal pressure changes over the first module, cancelling out the first action signal and exhausting that signal to atmosphere.

The second module, now energised, sends out its action signal. At the same time it sends a pressure signal back to the first module which locks the first module in the "off" position.

In this way, each step of the cycle is carried out until the last step has been completed. The feedback signal after the last step cancels out the last action signal and also cancels the locking pressure which held all the preceding modules in the "off" position.

The system is then ready for the start of the next cycle. The start may be brought about either by pressure maintained on the starting port of the first module or by the pulse signal from a manually operated start button.

The modules themselves have three parts which are fixed together with screws after which the modules are clipped on to a common mounting frame. The three parts are:

1. a sub-base with appropriate ports to match both its own main operating unit and its adjacent sub-base when clipped on to the mounting frame.
2. a main operating unit in which is the valve assembly—differential poppet type for small movement, large flow and fast action.

3. an attachment which converts the feedback signal into an acceptable form to change over the operating unit to which it is attached. Signals may be of pressure applied, pressure relieved, low pressure from proximity sensors, solenoid pilot receiving electrical signals, and so on.

As each module is clipped into place, ports in the sub-base match up each with the adjacent sub-base so that the signals from one to the other pass through without the need for external pipe work, other than for the action and feedback signals together with start and final signals to cancel the system ready for the next cycle. Figure 5.10 shows the symbol adopted by a manufacturer of such modules for the basic module. Figure 5.11 shows a simple circuit employing the modular program sequencer method controlling the operation of two cylinders with a cycle of operation according to the sequence shown.

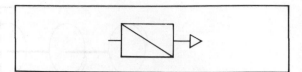

Fig. 5.10 *Manufacturer's symbol for modular program sequencer*

As Figure 5.11 shows, a module is used for each successive step. Each module, having sent out its action signal, is cancelled out progressively through the cycle. As a means of comparison with the alternative cascade system, the same sequence is shown in Figure 5.12. In this case, because the sequence of operation falls into a two-group system, only one five-port logic valve is employed as compared with the four sequence modules and the two end modules—six items in all.

Naturally in the initial investigation all these aspects would require consideration.

Fig. 5.11 *Simple circuit showing use of modular program sequencer*

Sequence
A +
B +
B −
A −

start

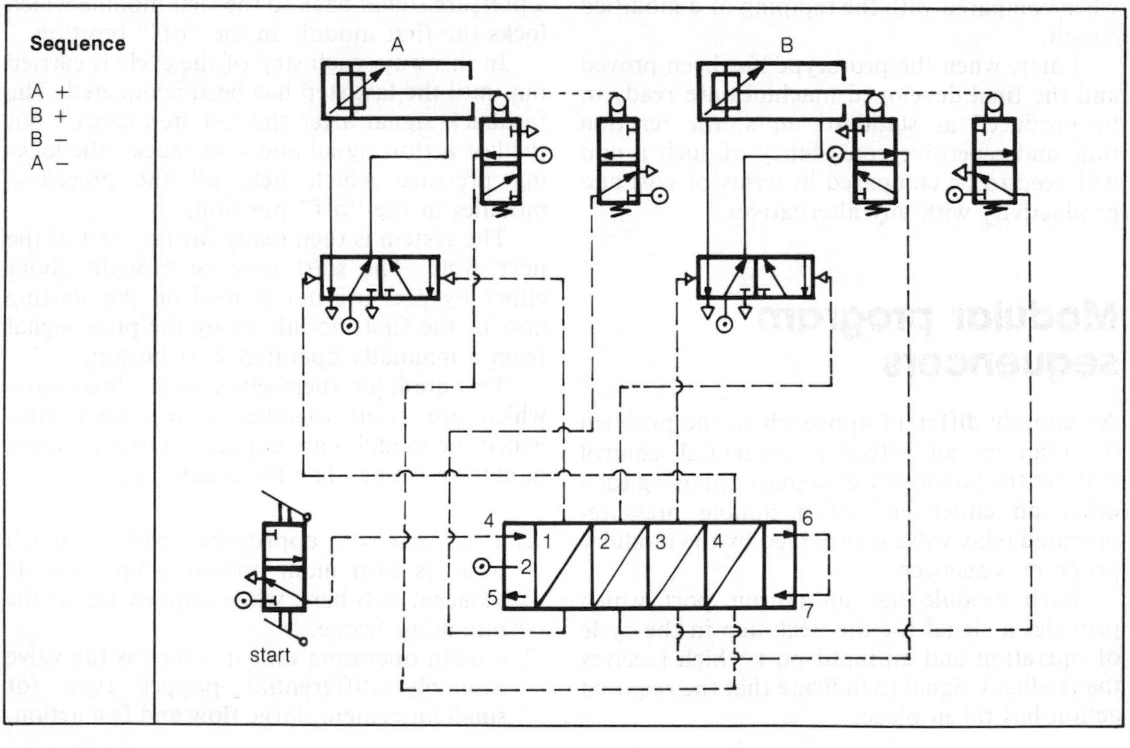

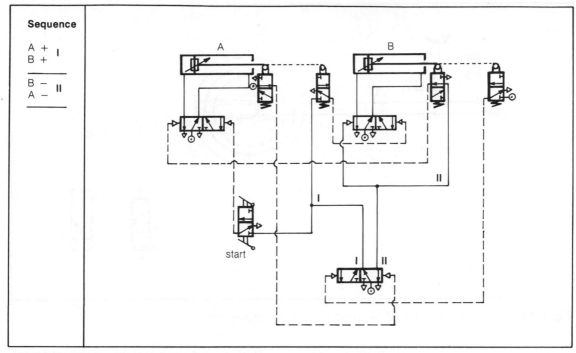

Fig. 5.12 *Cascaded circuit of same sequence as that in Fig. 5.11*

The modular method shown in Figure 5.10, with its ease of changing the program by reconnecting the feedback and action signals to appropriate modules, lends itself to the type of developing prototype project work where, as the project progresses, minor changes in the sequence may have to take place. Such changes would present no problems of a major repiping nature for the control circuit.

Each main module normally has a visual on/off indicator to aid diagnosis of unscheduled stoppages.

Bi-selector program sequencers

The bi-selector represents still another approach to the question of providing a control system which can positively produce a predetermined sequence of operation without running into problems of the "unwanted" signal, or opposing signals. Figure 5.13 shows in sketch form the functional concept of this type of unit. As the sketch shows, a centrepiece rotates through twenty stations. At each station there are two outputs—one a low pressure in the order of 80–100 kPa while the high-pressure output can be varied from 300–550 kPa for normal working.

The low-pressure output provides the feedback signals. When occluded at the orifice through which it escapes to atmosphere, there is a build-up of pressure back to the unit. When this pressure has built up to something in the order of 17–18 kPa, the centrepiece indexes to the next station.

On arrival at the next station, a new pair of outputs is emitted and the previous pair exhausted to atmosphere. Air pressure is emitted at only one station at a time. All the others are in a state exhausting to atmosphere.

The high-pressure output may be used either as a signal to change over a relay for action or

67

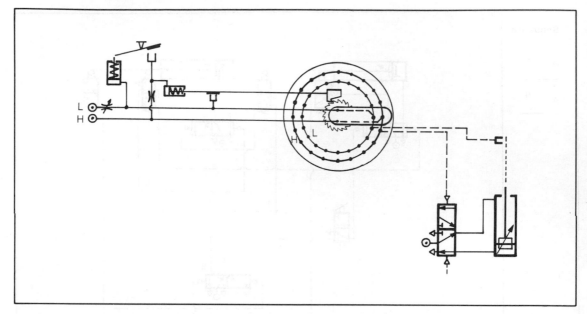

Fig. 5.13 *Functional concept of bi-selector program sequencer*

as the power to operate a small single-acting cylinder. This latter application is particularly suited to the programming of multiple stops, as, for example, in bending and forming sheet metal.

Occlusion of a low-pressure output causes the unit to index to the next station. Thus, any stations not required for the cycle of operation may be skipped simply by occluding the unused outputs. The unit will then index automatically from station to station until it reaches one with the low-pressure output free to atmosphere. There it will stop and both output signals, high and low pressure, will be given out.

Reliance on the feedback signal to index to the next station ensures that no step in the cycle will start until the preceding step has been completed. At the same time, programming is simply a matter of connecting up to the stations in the sequential order of the cycle. Changes in the program can be made either by altering the connections or using a manual change multi-NOT gate. With all connections to the bi-selector passing through such a device, manual operation of plungers will alter the program without disturbing the pipe work.

With a single unit catering for twenty steps in a cycle, such a device offers a method of control which is physically compact and economical in the use of components. At the same time, it is capable of controlling highly complex processes. It is not uncommon to find several such devices interconnected at the heart of sophisticated, complex processes.

Circuit symbol and drawing

Figure 5.14 shows the symbol used to designate a bi-selector and also the normal method of drawing a circuit when using it.

Because of the large number of signals involved, it becomes easier to read a circuit diagram if the destinations of the signals are shown as they leave the unit and these destinations are in turn identified with the same code.

As the diagram shows, a capital letter indicates a movement or an event in the cycle—a cylinder which will move, a valve to be turned on, etc. A lower case letter (*a*, *b*, *c*, etc.) indicates the sensor used at the completion of the step called for. Thus, *A* + indicates that the signal applied to the relay valve controlling cylinder *A* will cause cylinder *A* to extend. When *A* has extended, the sensor which shows that the outstroke of *A* has been completed will be designated as *a* +. This method of description and circuit drawing is often used when any program device has a large number of steps to be catered for in a cycle. Omitting the connecting lines from the drawing leaves room for the other components which must be shown, if they are part of the circuit. Such other components may be pressure regulators or, perhaps, small subcircuits using small logic valves additional to the sequencers.

As is shown in Figure 5.14, an AND condition is easily catered for when the signal used as a feedback to the device is obtained from the occluding of an escape orifice for the low-pressure output. Several escape orifices

may be connected to the one low-pressure output. All must be occluded before the bi-selector will index to the next station.

This attribute provides an economical and simple means of establishing a number of conditions as existing before the next step in the cycle takes place. For example, in positioning a component precisely before drilling or shaping, several small low-pressure jets may be used in appropriate positions, all from the one output. Or again, in packaging it may be necessary to establish that all six cans are in place before they are pushed into a carton. Six low-pressure jets, all from the one output, will establish that all are in place before the high-pressure signal for action for the next step can come from the bi-selector.

As with the program sequencer discussed earlier, visual indication is provided of the bi-selector's state in the event of an unscheduled stoppage. Diagnosis of the stoppage is consequently quick and easy.

Fig. 5.14 *Practical example of bi-selector circuit*

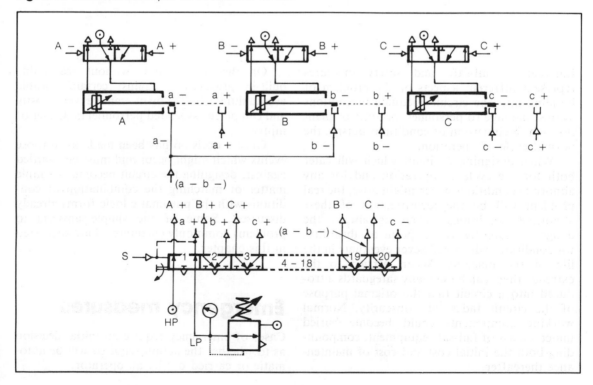

Chapter 6

Emergency, fail-safe and safety measures

Emergency, fail-safe and safety measures represent different aspects of the problem of building into the circuit adequate precautions against damage to personnel, product or plant through the intrusion of conditions outside the normal cycle of operation.

When designing a circuit which will cater both for the cycle of operation and for any abnormal conditions which might arise, the real problem will be the identification of these abnormal conditions. There is always the danger of catering, to the point of absurdity, for conditions which will never eventuate in the life of the machine. When carried to the extreme, there can be so many safeguards introduced into a circuit that the original purpose of the circuit fades into obscurity. Normal working components could become buried under a mass of fail-safe equipment, compounding both the initial cost and cost of maintenance thereafter.

On the other hand, without reasonable built-in precautions against certain feasible eventualities, breakdowns can be very costly and can place associated personnel in danger of injury.

Once a decision has been made as to those events which might occur and must be guarded against, designing the circuit becomes a simple matter of matching the combination of conditions with the pneumatic logic forms already discussed. Some of the simple answers to problems commonly encountered are suggested in this chapter.

Emergency measures

Cases of emergency require an initial decision as to whether the action required will be automatic or carried out by an operator.

Automatic response to emergency conditions

If the response to emergency conditions is to be automatic, the emergency conditions themselves need to be clearly identified and defined. They should then be written down. There may be several sets of conditions to consider. Listing them all ensures none is overlooked.

A typical example of the variety of effects resulting from a single change of a number of normal conditions is provided by a process involving the use of a conveyor belt. In this case, a hopper gate is normally opened by an air cylinder. The material then flows on to a conveyor belt driven by an electric motor. The conveyor belt transfers the material up to a higher level.

A failure in the electric power supply could cause several problems. What, for example, is likely to happen to the conveyor belt itself with its load? If no provision for the automatic application of a brake is made, the weight of the material will reverse the belt and bring the load on the belt back again to the hopper gate. If no provision is made to close the hopper gate when the power fails, the material will continue to flow. If no material flows from the hopper to the higher level, the process may be seriously affected at later stages.

Each of these possibilities needs detailed consideration and appropriate design action.

Operator response to emergency conditions

The two most difficult problems to solve associated with operator response to emergency conditions are:

1. the means of indicating that emergency conditions exist, and
2. the operator's first task as reaction to response.

Both problems will require close study of the working environment and the operator as a normal human being in a state of near panic.

Indication of the emergency state could be visual or auditory. If visual, is it possible to employ a means which will be so distinct from anything else in the vicinity that there can be no mistaking the indicator for what it is? If auditory, what sorts of sounds are already in the vicinity? Is another distinctly different sound possible?

So far as the immediate reaction of the operator is concerned, it is always well to require only the simplest possible action. If a number of tasks must be carried out, the first task should coincide with a natural, instinctive operator reaction. It should also, as far as possible, render the situation safe for the moment, giving time for the rest of the emergency tasks to be carried out in a more leisurely manner.

As a rule, it will be found that it is easier and more instinctive to push or bang down an emergency button than it is to pull or move over a lever.

If an emergency button is selected to deal with the situation, it must be placed where it is not only prominent but also easily reached. Its colour, shape and size should all combine to make it stand out against anything else in the vicinity.

Mechanical action

Usually, an emergency situation calls for an immediate stopping of all movement in the cycle. However, whatever action is decided upon the decisions will rely on full consideration of the overall cycle and the repercussions likely from any immediate stopping at any stage. Often, a cycle will include a particular movement which may be reckoned as potentially dangerous. Such a movement may be required, as soon as the cycle is stopped, to return to its normal position of rest. Again, it may be necessary either to return several movements, automatically or after a reset button has been pressed. Again, it may be more desirable for the cycle to stop for the emergency then, after remedial action has been completed, allow the cycle to carry on from where it stopped.

Stopping a cycle in an emergency

Exhausting the system

Figure 6.1 shows the use of a three-port valve placed immediately after the lubricator in the entry line of the main air supply to an application. Using a push/pull type it may be piped up so that the pushing of the plunger cuts off all air entry and exhausts all pressure from the system downstream.

Exhausting all cylinders as well as the pilot system can be dangerous. Sudden removal of all pressure may result in movements already in progress being carried on to the end of the stroke by the momentum of the mass in motion normally controlled by the cylinder. Thus a particular movement which may be expected to stop may not do so immediately. If the cylinder happened to be lifting a load, exhausting the pressure would naturally allow the load to fall.

Also, when air pressure is restored to the system, after remedial action has been completed, those movements which had not completed strokes when all pressure was removed would complete the strokes in an uncontrolled manner. Since none would have any exhaust pressure in front of the piston, they

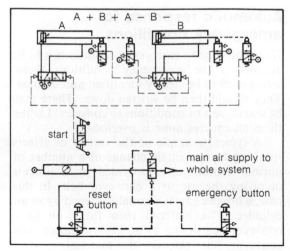

Fig. 6.2 *Exhausting whole system by remote control*

would accelerate to the end of their respective strokes with no means of cushioning the stroke at the end. This could result in mechanical damage.

Because of these associated repercussions following complete exhausting of the system, this method is rarely used. If used at all, it will

Fig. 6.1 *Exhausting whole system by direct manual control*

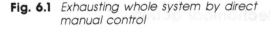
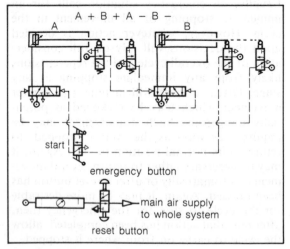

Fig. 6.3 *Direct manual exhausting of pilot system*

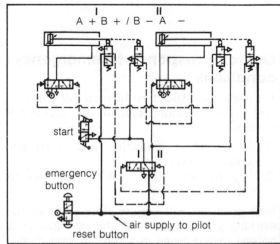

be used only for some very good reason. Figure 6.2 shows this method carried out remotely with a reset button to be used when pressure is to be restored again.

Removal of pilot air supply

A method usually found to be preferable to that shown in Figures 6.1 and 6.2 is that in which the air supply to the pilot system is removed and the pilot system exhausted to atmosphere.

Figure 6.3 shows this method using the direct control concept. Usually the component employed is a push button, air reset or pull return pneumatic switch, manually operated as required.

Figure 6.4 shows the same method controlled remotely from a signal which may be provided either by a valve depressed by the operator or by a pilot valve designed to provide a signal in response to a prescribed set of emergency conditions.

By substituting a five-port valve, in place of the three-port valve shown in Figures 6.3 and 6.4, an emergency pilot supply may be made available to any emergency subcircuitry required to complete emergency action. Figure 6.5 shows this type of circuit. As will be seen, it is commonly used in fail-safe circuitry.

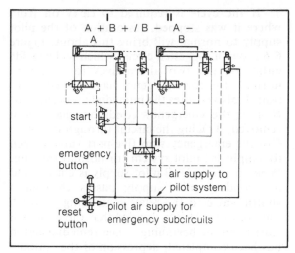

Fig. 6.5 *Remote exhausting of pilot system with simultaneous provision of supply air for emergency subcircuitry*

Exhausting the pilot system to atmosphere obviously will not stop a movement in progress at the time of the emergency action. That particular movement will complete its stroke. However, there will be no further movement after that since all sensors will have had their air supply removed. When the fault has been remedied and air restored to the system, all power units will still be controllable.

After the cycle has been stopped, further action is often required before the process can return to normal operation. The circuit shown in Figure 6.5 provides a typical example of the sort of action required.

It could be that movement *B* should always retract when anything goes wrong with the operation. Then, when the operator has cleared the trouble, either the whole system should return to the start position or the cycle may carry on from the point in the cycle at which it was stopped by the depressing of the emergency button.

If the cycle is required to return to the start position, a reset button should be included which will return the pilot supply to line II. This would result in cylinders *A* and *B* returning in their correct sequence.

Fig. 6.4 *Remote exhausting of pilot system*

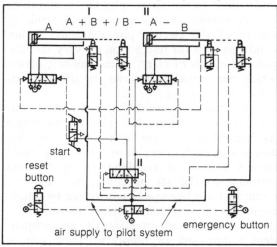

If the cycle is required to carry on from where it was stopped, restoration of the pilot supply to normal will bring this about. Figure 6.6 shows a suggested circuit which will provide either of the two alternatives to return to normal after the emergency has received attention. It also ensures the immediate retraction of *B* until the emergency conditions have been removed. Tracing the circuit through in Figure 6.6, the emergency stop five-port valve is seen to supply the pilot system for normal running. When depressed, all pilot supply is cut off and a secondary pilot supply passes through a shuttle valve to the relay controlling *B* which then retracts. If it is decided that the cycle will start from the beginning, when remedial action has been completed, depression of the optional reset changes over the pilot supply valve to line II. When the final reset is depressed and normal pilot supply restored, *A* will retract after *B* and the system is then ready for the start switch to be switched on to resume normal running. It should be noted that, as soon as the emergency stop has been depressed, the start switch should be turned off until normal running is resumed.

Fail-safe measures

Fail-safe measures are usually understood to be those safeguards which are included in a circuit to cover the intrusion of certain foreseen possibilities which may occur outside the course of normal operation. When those undesirable combinations of conditions occur, remedial action to render the system safe is designed to take place automatically.

Once the likely undesirable conditions against which precautions must be taken have been defined, the methods of sensing these conditions will require careful consideration. The components selected to carry out any action will need to be matched against the worst possible conditions to ensure that their characteristics are capable of producing a reliable response. For instance, not every pressure switch can be relied upon to react to a rise in pressure at precisely the same point every time. Not every pressure switch will produce a

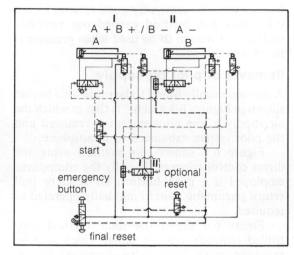

Fig. 6.6 *Emergency exhausting pilot supply with subcircuit to retract B. Reset to normal start*

positive signal when sensing a gradual change of pressure. Valves or pressure switches employing springs will need to be examined to ensure they will change over with a snap action. Reservoirs storing emergency air supplies will need to be leakproof in all respects—fittings, valve components, etc.

The two most common conditions which call for implementation of fail-safe action are probably failure of the electric power supply and failure of the air supply. Often failure of the air supply follows as a natural result of failure of the electric power supply. Either condition calls for reliance on the very simplest possible means of effecting the fail-safe action which automatically follows the detection of the need for action.

The means of initiating fail-safe action, then, must rely as much as possible on simple mechanical devices employing counterweights and springs rather than air or solenoid operated devices. The action resulting from the initiating response to the set of conditions requiring attention must likewise be carried out by simple means. If the emergency action involves cylinder movement and valve signals, an independent compressed air supply for these will be required. A reservoir, large enough for the purpose, must therefore be provided.

Figure 6.7 shows a typical fail-safe circuit providing for the outstroking of a particular cylinder in the event of a drop in pressure in the main air supply below a predetermined point. The main supply pressure is sensed continuously by a pressure switch. The pressure switch is set to pass a pressure signal from the reservoir when the pressure of the main supply drops below the level set for the switch. The emergency signal changes over the two three-port valves. The three-port valve connected to the front of the cylinder allows any pressure in front of the cylinder's piston to exhaust to atmosphere. The three-port valve connected to the rear of the cylinder changes over to connect the reservoir to the rear cylinder entry port. The reserve supply in the reservoir then expands to drive the piston forward.

As may be seen, it is essential that the reservoir is large enough to allow its stored air to expand to the volume required by the reservoir itself plus the volume of the air cylinder and the volume of the signal lines. In expanding, the pressure must not drop below the safe working pressure for both the cylinder and any pilot valves used in the fail-safe system.

Fig. 6.7 *Fail-safe circuit with automatic outstroking of cylinder*

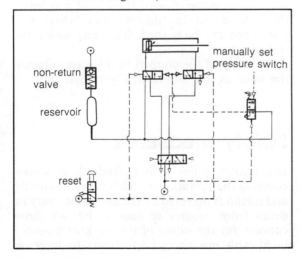

Method of sizing reservoir

The method employed to assess the volume required for the reservoir relies on the simple formula representing Boyle's Law: $P_1V_1 = P_2V_2$. Taking Figure 6.7 as a working example, let the cylinder shown there represent one having a bore diameter of 100 mm and a stroke of 200 mm.

Then, V_1 will represent the unknown volume of the reservoir.

V_2 will represent the reservoir volume plus that of the swept volume of the air cylinder.

P_1 will represent the normal minimum main supply pressure—550 kPa in terms of absolute pressure—650 kPa (abs).

P_2 will represent the minimum safe working pressure for the cylinder and valves requiring action in response to the emergency fail-safe signal, 450 kPa or 550 kPa (abs).

Extending this further:

V_1 = reservoir volume = R
V_2 = R + swept volume of cylinder
= R + $\dfrac{\pi \times D^2 \times \text{stroke}}{4}$

From Boyle's Law:
$P_1V_1 = P_2V_2$

Substituting with actual figures:

$650R = 550 \times (R + \dfrac{22}{7} \times \dfrac{100^2}{4} \times 200)$

$\therefore \dfrac{650R}{550} = R + \dfrac{22}{7} \times \dfrac{100^2}{7} \times 200$

$\therefore 1.1818R = R + 1\,571\,428.5$ mm³
$= R + 1.571\,428\,5$ L

$\therefore \quad R = \dfrac{1.571\,428\,5}{0.181\,8}$
$= 8.64$ L

Since a small allowance is necessary for piloting fail-safe circuit valves and also to cover small possible losses through leakage, the figure shown here would be increased to a nominal figure of, say, 10 litres.

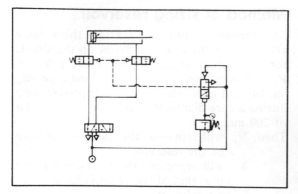

Fig. 6.8 *Emergency locking of cylinder*

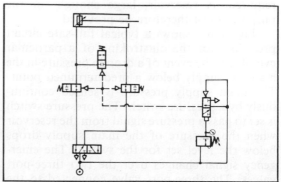

Fig. 6.9 *Cylinder in emergency free to move manually*

Locking a cylinder in position

There may be occasions when, instead of a particular cylinder extending or retracting in the fail-safe condition, a cylinder is required to lock in its position at the time of the emergency. This can be achieved by cutting off all entry of air and exit of exhaust by means of poppet valves. These, under normal conditions, should be held open by pressure emanating either from a pressure switch, pressure-sensitive valve or a differential pilot valve. The pilot valve should be set to retain the signal pressure on the poppet valves until mains pressure drops below a pre-determined level. Exhausting of the pressure ends of the poppet valves allows these to close. Cylinders locked in this manner can rarely be relied upon to remain so for an indefinite period as any slight seepage of pressurised air from either end of the cylinder will cancel out the locking effect. The method, however, is often useful as an emergency or a fail-safe measure when such a state of emergency is known to call for positive immediate remedial action on the part of maintenance staff. Figure 6.8 shows a typical circuit employing the method just described.

Cylinder free to move manually without exhausting

Where a state of emergency calls for the force to be removed from a cylinder so that it may be physically manipulated one way or the other in its stroke, a useful method to employ is shown in Figure 6.9. This uses two locking poppets open and one inter-flow poppet closed when pressurised.

Exhausting the air in the cylinder will achieve the same effect as that shown in Figure 6.9. However, when air pressure is restored to the system, after remedial action, any cylinder exhausted to atmosphere presents problems described earlier. With no exhaust pressure, the movement such a cylinder may make to restore its previous position will be uncontrolled. No exhaust in front of the moving piston will result in possible violent, non-cushioned movement which can cause mechanical damage.

On the other hand, allowing free flow of pressure to either side of the piston will leave the cylinder in the desired state where the piston rod can be pushed either way with little resistance.

Piloting of the poppets in this case is much the same as that shown in Figure 6.8.

Safety measures

Every country has certain standards of safety covering the operation of machinery. Standards and methods accepted by the authorities vary in detail from country to country but all show concern for the safety of the workforce associated with machinery. As efforts to increase

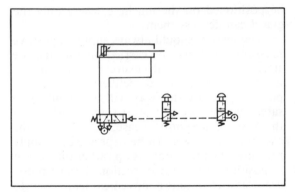

Fig. 6.10 *Two button safety start of doubtful value as one button can be tied down by operator*

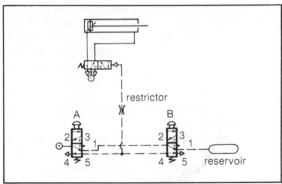

Fig. 6.11 *Two push-button safety start where both must be depressed simultaneously*

production rates have intensified and operators have responded to incentives, safety methods, acceptable in the past, have had to be revised and improved.

Figure 6.10 shows what was a common method of attempting to avoid the risk of an operator damaging fingers or hands when pressing a start button at the beginning of each cycle of operation. The diagram shows an air supply fed to the inlet of one push-button valve, the outlet port of which is piped to the inlet port of a second push-button valve. The initiating signal for the cycle is delivered from the outlet port of the second push-button valve. Ostensibly, both buttons must be depressed to gain a starting signal. However, with operators on incentive bonus schemes, many found that instead of using one hand for each push-button greater speed could be gained by tying down one push-button. This left one hand free to feed the workpiece into the machine, or to clear the product from the machine after it had received its treatment. Damage to the free hand was common. Consequently, many countries no longer accept such a method as a safety feature.

In addition to considering the ease of cancelling out any safety feature, any method adopted must provide for the possibility of the presence of people other than the operator in the vicinity. If this can be ruled out as a possibility, the two-button method shown in Figure 6.11 is popular. In this method, both buttons

must be depressed simultaneously to give a start signal.

Referring to Figure 6.11, it will be seen that, in a position of rest, main air supply connected to port 2A passes through valve A to port 2B and thence through valve B and port 1B to a small reservoir. Ports 3A and 3B are plugged.

When A and B are depressed simultaneously, air in the reservoir passes through ports 1B and 4B. Port 5A is cut off leaving only one route for the air to take, namely, through the restrictor to the pressure end of the relay valve connected to the cylinder. The cylinder then extends. Release of either A or B will allow the pressure signal to exhaust to atmosphere through the valve released.

If A is depressed but not B, air in the reservoir exhausts through ports 1B, 2B, 1A and 4A to atmosphere. If B is depressed but not A, air in the reservoir exhausts through ports 1B, 4B, 5A and 4A to atmosphere.

If A is tied down the reservoir could never be charged with air. If B is tied down, again the reservoir could never be charged with air.

Because the signal is delivered by the air expanding from the reservoir into the signal line connecting the valve B to the relay, the size of the reservoir must be checked to ensure that, in expanding, the air does not drop its pressure below the working limits of the valve receiving the signal. As with the reservoir described earlier for emergency supply, the volume may be calculated by applying Boyle's Law, $P_1V_1 = P_2V_2$. In this case, V_1 will be the volume of the reservoir, V_2 will be the reservoir volume plus that required to fill the signal line and the displacement volume of the relay valve spool as it moves over. P_1 will be the normal working absolute pressure and P_2 the safe minimum working absolute pressure required to change over the relay valve.

A small additional allowance should be made in respect to the reservoir volume to cover possible leaks and also the slight initial pressure drop likely to result from the drop in temperature as the air expands into the signal line. Ten per cent should be sufficient as a margin for such contingencies.

Because of the possibilities of people other than the operator placing themselves in danger while a machine is working, many authorities insist on a guard of some type to shield potentially dangerous movement.

In the case of automatic operation, such as an automatic power press, the circuit must be drawn up to include a sensor which senses the presence of the guard in its correct position while working and also provides the air supply to the pilot system which supplies the signals to keep the operation cycling. The circuit must be arranged so that, if the sensor ceases to supply a signal, the operation will stop. In the case of a power press, the clutch will disengage before the press reaches the end of its upstroke.

Figure 6.12 shows a suggested circuit for a power press. The sensors for the guard should

be positioned on the machine where only the guard can depress them.

The sensors should always supply a positive pressure which will be carried on through the start switch. This will ensure that any movement will be initiated deliberately. Use of positive pressure this way will be found to be safer than the use of a start signal working on the basis of pressure release, as might be the case if a sensor were to be depressed to supply pressure when the guard is up rather than down in position, ready for operation. Use of pressure release to initiate a movement lends itself to release of pressure through some other means, such as a loose fitting in the pipe connections.

In the case of a machine with a number of movements, one of which must return to retracted position if the guard is lifted, the circuit shown in Figure 6.5 may form the basis of a circuit to provide such action on the lifting of the guard.

Fig. 6.13 *Safety start circuit where all movement ceases if guard lifted*

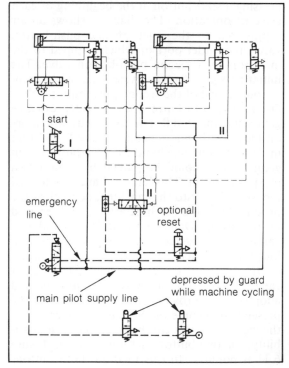

Fig. 6.12 *Two push-button start with safety guard sensor incorporated*

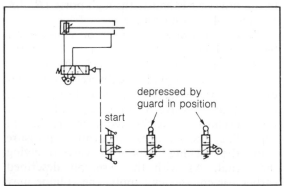

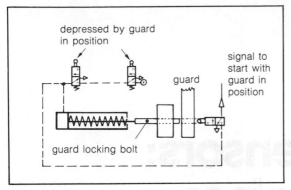

Fig. 6.14 *Safety start with guard locked in position by air cylinder*

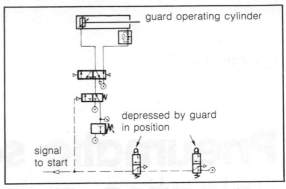

Fig. 6.15 *Safety start with low pressure air in cylinder until guard automatically positioned, then guard held by cylinder with high pressure air*

Figure 6.13 shows a circuit incorporating that of Figure 6.5 with sensors ensuring that the guard is in place while the machine is operating and that all movement will stop if the guard is lifted together with retraction of a particular cylinder.

Some authorities insist on the guard being locked in position before a machine may start working. A common method employed to meet such requirements is shown in Figure 6.14. In this case, when the guard is in position, a sensor causes a small single-acting cylinder to extend. In doing so it pushes a locking bolt into place. The locking bolt, in turn, depresses a sensor which supplies a signal to start the cycle. As with previous sensors described, the sensors need to be placed in positions where they may be operated only by the means designed into the machine.

Bringing a guard into position as the first movement in an automatic cycle can make the guard itself a hazard if it is brought into place with too much force. Figure 6.15 shows one way of ensuring that the guard is automatically brought into place by an air cylinder in which the thrust is reduced to a level unlikely to damage a hand or fingers if it comes into contact with them. The cylinder movement outstroking works on a safe pressure and speed. When the cylinder has outstroked to place the guard in its working position, the sensors serve to indicate that the guard is in place, increase the pressure in the cylinder to hold the guard firmly in place and also deliver the start signal for the main operation to start cycling.

The suggestions put forward to meet emergency, fail-safe and safety requirements are but typical of means already commonly used. There will be many alternatives suggested by experience and imagination of the individual designing the project. Whatever method is adopted, it is important that it be the product of as intimate a knowledge as is possible of the working environment of the proposed machine and the people who will be expected to use it.

Chapter 7

Pneumatic sensors:
Selection guidelines

Pneumatic sensors are used predominantly to detect physical presence or absence and changing pressure conditions which may be the direct or end result of other related changes. A superficial review of pneumatic sensors may give the impression that they cater for only a small proportion of the sensing devices required in modern industry. Any such impression is soon dispelled by a little imaginative speculation on the application of sensors which will react to predetermined pressure states.

Some common applications of sensors reacting to pressure change are:

1. sensing the end of a cylinder stroke by the decay of the exhaust pressure in front of the piston.
2. sensing the level of a liquid by the change in pressure of the air trapped in a pipe immersed in the liquid. Rising liquid level compresses the trapped air, falling liquid level allows the trapped air to drop in pressure.
3. sensing the proximity of an object by the change in pressure upstream from a low-pressure jet when the jet flow is restricted by the close proximity of the object to the jet nozzle.
4. sensing the change of pressure created upstream from a low-pressure jet when the nozzle is occluded.
5. sensing the change of pressure upstream from a receiver nozzle by the interruption of a low-pressure jet directed across a gap at the receiver nozzle.

These are just a few common examples of the use of changing pressure as a simple means of determining when a change of condition has taken place. All of them provide simple, highly sensitive sensors which will perform reliably under working conditions which are often unacceptable for the equivalents using other control media.

Because of their simplicity, the design and operation of pneumatic sensors can be readily understood by maintenance staff after minimal instruction. This does much to keep "down time" and maintenance costs to a low level.

To meet the needs of industry there are many different types of sensors from which to choose. They differ from each other widely in respect to robustness of construction, sensitivity, repetitive consistency and modus operandi.

Selection of sensors for any project is not a haphazard matter of reaching out for the nearest and cheapest. Careless selection can prove costly later. Nor is the cheapest in the first purchase cost always the cheapest in the long run. Rarely is this so. The imagined saving in the first cost can be lost many times over in three particular ways after the machine has been put into service.

1. Poor quality control of the product processed by the machine can be the direct result of components inherently inconsistent in their repetitive performance.

    Poor quality control will also result from the use of sensors incompatible in their sensitivity with the fine tolerances required by the product.

    Substandard products caused by poor selection of sensors can, in a very short term, destroy any chance of survival the product may have enjoyed in a competitive market.

2. Frequent malfunctioning of inferior, cheaper sensors can result in losses of production while repairs are made to the offending components and the machine to which they are fitted. These production losses, or down time, can amount to astonishingly high figures when converted into cash. If the machine is rated to produce a specific quantity of products per hour, for every hour it does not do so the turnover of the factory is reduced by the value of the product which has not been produced. This,

of course, is a direct loss in cash to the industry. One such unscheduled stoppage of the machine will cost far more, as a rule, than any difference in cost between the right component for the task and one which seemed to be cheaper.

3. Losses of production can needlessly occur through the down time required to make frequent replacements of components with a shorter life expectancy than those with a slightly higher first cost.

These aspects are not always fully comprehended by the conscientious purchasing officer in industry. Consequently the design engineer must have sufficient confidence in the reasons for the final selection of components to stand firm on the choice and justify it by explaining those reasons.

It should always be remembered that those who question the economics of the project or the price of the components selected are just as interested in ensuring that the industry advances on solid financial ground as is the engineer in ensuring that the industry advances on solid technical ground. The two are complementary. Therefore, the reasons for final selection of components must be translated into terms of money lost or gained through loss or otherwise of quality control and production time. Interdisciplinary respect and cooperation will provide the means of attaining a reliable and effective machine in a sure and lasting manner.

The variety and range of sensors is so extensive that its very profusion can make it a difficult matter to select the correct component for the particular task under review. However, such a profusion can be rationalised and the selection made easier by classifying sensors as they conform to the characteristics of one or other of a number of general types. Each of these types can be related to a particular design feature. With sensors grouped in this way the general characteristics of each group can be clearly defined. It then becomes a relatively easy matter to match the functional requirements and environmental working conditions with a particular component.

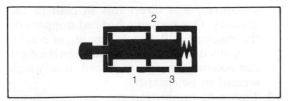

Fig. 7.1 *Plunger-operated spring return, spool type sensor*

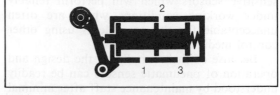

Fig. 7.3 *Roller-operated spring return spool type sensor*

Sensor employing external and internal mechanical operation

Design

This type requires physical contact from a cam mounted on the apparatus or device whose presence or absence is to be sensed and indicated by a pressure signal transmitted from its outlet port. The cam or contact surface of the device to be sensed makes contact and exerts physical pressure on a plunger, either directly or through a mechanical linkage incorporated in the mechanism of the sensor. For example, a roller, mounted on the end of a lever attached to a pivot point, may be depressed, causing the lever to depress a plunger.

Figure 7.1 is a sketch of a typical arrangement of a plunger-operated sensor. Also, shown in the same sketch is the mechanical internal arrangement required to return the sensor to its original state after the plunger is released. In this case a spring is used. The mechanism also shows a spool as the device which changes the direction of air flow through the sensor—inlet to outlet, or outlet to exhaust port.

Fig. 7.2 *Plunger-operated spring return poppet type sensor*

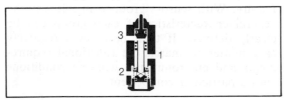

Figure 7.2 shows an external mechanism operating an internal mechanical arrangement using a poppet valve, directional flow changing method. There are several different versions of the poppet valve approach in common use.

Figure 7.3 shows a typical mechanical external arrangement used to sense a moving cam. The cam makes contact with the roller. When the roller is depressed, it causes the lever arm to which it is attached to depress the plunger, which directly changes over the component directional flow mechanism.

General characteristics and useful guidelines

Because these sensors sense the actual physical presence of the device to be sensed, they provide positive evidence. Thus they are widely used in all kinds of industry—heavy and light. They vary in construction from heavily built, rugged types requiring relatively considerable force to operate them to lightly constructed, miniaturised components requiring very little force and very little movement.

Materials used in the construction of these components include brass, steel, aluminium alloy and various plastics. These materials must be matched against the working environmental conditions of the component under review.

A typical example of difficult working conditions for sensing components is a cheese factory. There, all working machinery is in direct contact for long working periods with water, steam, lactic acid from the cheese and caustic soda in the daily washing down water. In matching the materials against such corrosive elements, it is easy to overlook some small but vital item. For instance, gunmetal, stainless steel and certain types of plastic materials may have been used in such an exercise for the obvious parts which go to make up the whole sensor with the exception of, say, the circlips which may hold the roller axle on the end of the roller arm or lever. Or again, the screws holding the operating mechanism to the sensor valve body may have been overlooked. If both, or either, of these items are made of the normal spring or mild steel, they will rust through in a few weeks. As with the other components, circlips and screws must be of rustless materials, usually stainless steel. If these small items are overlooked, the sensors can cause numerous stoppages while they are replaced. In this, as in any other enterprise, the chain is only as strong as its weakest link.

Obviously, materials are important and require close study in matching in every detail with the working environmental conditions.

Mechanical movement required to operate the sensor requires consideration from several different aspects.

Cam design, for the cam which will move over and depress a roller operating mechanism, must provide an angle of approach no greater than 30° from the line of approach. Where the cam is to depress a sensor while moving over the sensor, the design of the cam must take into account the speed of the moving cam as related to the distance along which the sensor needs to be held depressed to provide a signal. Consideration of the signal to be given by the sensor will entail checking the length of signal line to the component receiving the signal and also the response time of the recipient of the signal.

Thus, before the operating length of the cam may be specified, the following facts must be assessed and assembled:

1. maximum anticipated speed of moving cam,
2. Cv factor of sensor selected,
3. pilot pressure to be used,
4. length and bore of signal line,
5. pilot-operating pressure required by recipient valve,
6. response time of recipient valve.

Using the methods described in Chapter 8, the length of signal time required from the sensor may be estimated. With the signal time established, it must then be related to the maximum anticipated speed of the moving cam. This will establish the required length of the operating portion of the cam, i.e. the flat portion of the cam which rides over the roller while holding the roller mechanism fully depressed.

Example

Estimated length of signal time required 0.06 second

Speed of moving cam 0.5 metre per second
Therefore,

Cam operating surface length

$$= \frac{500}{1} \times \frac{6}{100} \text{ mm}$$

$$= 30 \text{ mm}$$

Add a minimum of 10% to cover variables so that final length will be 33 mm

Figure 7.4 shows in sketch form what is meant by "operating surface" (*X*) and "angle of approach" when applied to a cam.

The operating movement required by a sensor needs to be reviewed to ensure it is compatible with the speed and accuracy or precision required from the project when it goes into production. The length of operating movement is usually a means of making a preliminary general assessment of the sensor's performance characteristics.

Fig. 7.4 *Mechanical cam arrangement*

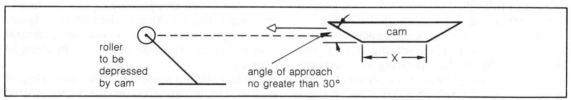

It is customary to find operating movement details provided by the manufacturer and stated as in the following examples:

1. Pre-travel 2.4 mm
   Operating movement 1.5 mm
   Total movement including over-travel
   5.5 mm
2. Pre-travel 0.8 mm
   Operating travel 0.8 mm
   Over-travel 1.5 mm
3. Pre-travel 3.5 mm
   Operating movement 4.5 mm
   Over-travel 1.5 mm

All three examples are sensors which are mechanically operated from cams and employ internal mechanical means of changing over the flow direction.

It will be noted that the operating movement in the first two examples, 1.5 mm and 0.8 mm, are both considerably less than that of the third example, 4.5 mm. As will be seen when sensors using other than internal mechanical means to provide a signal are examined, the first two examples require considerably greater movement than some of the other types available. Nevertheless, they require a good deal less operating movement than the third example. The reason for the marked difference between the third example and the other two is simply that the internal mechanical arrangement is different. The first two employ a poppet valve arrangement and the third employs a spool. From a practical point of view, the poppet and spool arrangements both produce slight, but important, differences in the characteristics associated with sensors using either one arrangement or the other.

Spool sensors normally require greater operating movement than any other sensor of the physical contact type. Because of this changeover time is relatively long, both in respect to full signal on from no signal and full signal off to full exhausting of the signal.

Accordingly it is not good practice to use a spool type sensor when sensing a very slow movement. With the operating cam creeping over the operating roller of the sensor, the slow spool movement produces a signal which slowly builds up pressure. The recipient valve, subjected to such a slow build-up, will often start to change over. As the changeover mechanism meets varying resistance in its changeover action, it may stop moving until the signal pressure has built up to the higher pressure required to move the changeover mechanism further. Certain types of relay valves, at this stage, can stop long enough to allow their pressurised seals to be extended into the ports across which they must pass. By the time signal pressure has built up to its maximum, the effort to move the changeover mechanism is beyond that which the signal can provide. The recipient valve may then stick in mid position. Thus the slow build-up of no signal to full signal resulting from the slow moving of a spool type sensor can create problems.

Problems of another nature are created by the relatively long time taken for a spool sensor to revert from full signal to no signal, in which position the previous signal pressure is allowed to exhaust to atmosphere. This relatively long changeover time is, of course, due to the distance the spool must move, propelled by a spring. Even though the cam's angled approach surface is kept as short as possible and the cam is removed as quickly as possible, the spring return, in itself, can contribute to a relatively slow changeover. The spring, by reason of its inherent characteristics, exerts a diminishing thrust against the spool as it expands with the spool movement.

Slow return to exhausting position by the sensor creates problems of a far-reaching nature when high speed production is aimed at.

After a relay valve has been changed over to one position by a pressure signal, it cannot be changed back to its original state until its last pressure signal has been removed from the pressure end. Pressure from the second signal, which is intended to restore it to its original position, cannot move the changeover mechanism against such an opposing signal. The longer the time taken to exhaust the previous signal, the longer must such a double pressure-operated relay valve wait until it can be changed over again.

In fast cycling machines, slow exhausting of earlier signals can account for many wasted milliseconds on each relay valve. To the uninitiated a few milliseconds here or there may not seem to be of any concern. However, a couple of examples will suffice to show that productivity can be seriously affected by this.

A high speed machine with four basic movements, each following the other after the preceding movement has been completed, has for its sequence of operation:

$$A + A - B + B - C + C - D + D -$$

Such a cycle takes two seconds to complete. However, in each case the signal to instroke is received by each relay valve in turn when its outstroke signal pressure is still not completely exhausted. In each case no movement occurs until fifteen milliseconds have elapsed—the time taken for the outstroke signal to exhaust adequately. Thus a total of sixty milliseconds is wasted in the cycling time. If this waiting time were eliminated by using a sensor compatible with the machine's functional requirements, the cycling time would be reduced to a total of 1.94 seconds. In terms of productivity from that machine, correct selection of the sensors would produce an increase of 3.09 per cent production.

A second example of the effects of slow exhausting of signals through using the wrong type of sensor for the particular application was provided by some mechanical equipment designed for use in the meat industry. Part of the project relied on rapid oscillation of a device powered by an oscillating air cylinder.

The specification called for a double-acting, cushioned air cylinder to oscillate over a stroke of 350 mm at an overall average speed of 1.5 metres per second. The bore of the cylinder was 65 mm and the relay valve selected was of ½″ BSP porting and Cv factor of 3.3. Sensors initially applied were of ⅛″ BSP porting with a Cv factor of 0.14. To keep the apparatus as simple as possible, oscillation was controlled by a roller-operated/spring-return sensor at each end of the cylinder stroke.

Because of the slow exhausting of the sensors, the first trials produced these results:

1. The cylinder never attained desired speed; after the first four or five strokes loss of cushioning at each end of the stroke became apparent and the loss of cushioning increased as the stroking was allowed to continue until cushioning was non-existent.
2. After some twenty-five strokes, the cylinder lost power progressively, also speed diminished until the cylinder stopped with the relay valve jammed in mid position.

The sensors were then changed to a type requiring physical contact externally but using an internal arrangement of a combination of air relief and differential spool. This type was successful simply because of its faster response in its on and off changeover, combined with a small mechanical movement required to initiate its changeover action. The mechanical operating movement required by the first had been 4.5 mm, whereas the second, successful sensor required only 0.8 mm movement to initiate it in either state, on or off. As a result, using the second sensor, the relay valve controlling the cylinder was cleared of any residual exhaust at either pressure end by the time it had completed its stroke in either direction.

The slow exhausting by the first type of sensor used had resulted in the spool of the relay valve receiving the signals being subjected to what was akin to a diminishing differential of pressure applied intermittently to either end of the spool. As a consequence the spool gradually moved less and less in either direction, eventually finishing up in the mid position.

Poppet type sensors, size for size, usually have a smaller operating movement than their equivalent spool types. Because of this, combined with the fact that a poppet valve can produce a relatively large flow with a small valve seat movement, they produce the effects described as those associated with spool valves in respect to high speed and low speed to a lesser degree than produced by the spool type. Nevertheless, the tendency to produce the results described is there and should not be overlooked.

Where either type, spool or poppet, is used, it should be remembered that the greater the operating movement required, the more the opportunity that exists for variables to upset repetitive consistency of performance. Thus, if fine quality control is required and repetitive consistency of performance of the machine is important in processing the raw material through to the end product, external and internal mechanically operated sensors should either be avoided or used with care and discrimination.

On the other hand, sensors of this type are usually of a much more rugged construction than their more sensitive equivalents employing other modus operandi. They are often built to operate under conditions for which there could be no other types of sensor, pneumatic or otherwise, which could operate reliably for long periods of time. Accordingly, they have an exceedingly useful place in many heavy industries and other industries where the unavoidably difficult working environments render any controls, other than pneumatic, unreliable.

They are also used in light industry extensively where precision is less important than simplicity of construction and operation.

Displacement air escape or "bleed" holes

Close examination of the average roller-operated/spring-return sensor, especially of the spool type, will reveal a small hole at the base of the valve. This hole is there to allow the air, which is displaced by the spool moving down inside the valve, to escape. When the spool is returned to its original position by the internal spring, air flows back through the hole to fill the space no longer occupied by the spool as it moves upwards again.

If this hole becomes blocked with dirt, either one or other of two conditions can arise.

1. Resistance can be increased considerably to the depression of the spool, causing an increase of force required to do so.
2. Return of the spool by the spring will cause the formation of a vacuum effect which will resist the action of the spring sufficiently to prevent the full return of the spool. In turn, this prevents exhausting of the signal previously delivered by the sensor when it was depressed.

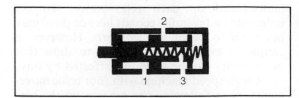

Fig. 7.5 *Spring-return arrangement exhausting spool displacement air through spool wall to exhaust port*

Under normal conditions, while the hole is open and allows free flow of air in either direction, the inflow of air, when the spool is being returned by the spring to the sensor's "off" position, often drags in contamination from the immediate outside vicinity of the sensor. For example, any liquids, corrosive or otherwise, around the base of the sensor are sucked in by the air inflow, to the detriment of the sensor concerned. Prevention of harmful effects of such bleed holes can be carried out in a simple manner.

The spool normally is hollow, in part, to allow space for the return spring. A small hole of equal diameter to the bleed hole in the base of the valve may be bored through the spool, between the two lower seals, from the outside of the spool to the inner spring recess. This hole, positioned between the two lower seals, will be found to be opposite the valve's exhaust port. Any displacement air can then flow through to the exhaust port of the valve and the bleed hole in the base may then be permanently blocked. Figure 7.5 shows the method described above.

Sensors employing external mechanical and internal pressure operation

Design

Figure 7.6 shows in sketch form the general internal arrangement of the means employed in changing over the state of the valve from "on"

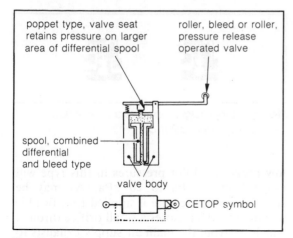

Fig. 7.6 *Roller/bleed sensor*

down in the "off" position. When the seal is removed, or lifted off the escape orifice, the pressure on the larger end exhausts through the orifice to atmosphere. Pressure on the smaller end then drives the spool upwards to change the valve to the "on" position. Closing of the orifice at the top causes pressure to build up again on the larger end of the spool, driving the spool down to the "off" position again.

Because the escape orifice at the top is larger than that of the longitudinal hole in the spool which supplies the air for this operation of the spool, air pressure cannot build up on the larger end of the spool until the exhaust orifice is closed by the seal.

Characteristics and useful guidelines

Small mechanical movement required to operate the valve

Because of the relatively large diameter of the exhaust orifice covered by a flat seated seal, the poppet characteristic of large flow with a small mechanical movement applies in this case. Many sensor valves of this type require only as little as 0.05 mm movement to initiate the internal pressure changes. The exhaust orifice has a larger area than that of the hole drilled longitudinally through the spool to supply air to both ends. However, the exhaust orifice is still small enough to require little effort to hold a seal against it in the closed position. Consequently, the effort required to operate such sensors is comparable with any of the most sensitive equivalent electrical microswitches.

Since the spool is changed over by bringing about a state of unbalance of forces by air pressure, rather than springs, the spool changes over with a snap action as soon as the state of unbalance occurs, one way or the other.

All of these design factors combine to provide a repetitive consistency of operation and fast response which is usually of a considerably higher level than may be expected from the first type of sensor discussed—namely those employing mechanical means internally and externally.

to "off" or vice versa. The external mechanical arrangement shown is only one of a number of arrangements commonly employed. This type uses a roller on the end of a lever. The roller may be depressed either by a cam riding over it or a cam which comes to rest on it at the end of a movement. Other external arrangements may use various forms of plungers to initiate the internal means of operation.

The internal operation is usually described as a "bleed" or "pressure release" method.

From the sketch in Figure 7.6 it can be seen that the valve employs a differential spool through which a longitudinal hole has been drilled. An air entry is provided at the base of the valve opposite the smaller end of the spool. The larger end of the spool has space in which to move and an air exit port at the top. This port is covered by a seal held in position by the mechanical operating external lever.

The spool is moved by pressure. With a constant air supply to the base, pressure is always applied to the smaller area of the bottom end of the spool. The same air supply also passes up through the spool to the larger, upper end of the spool. While the seal in the exit or escape orifice is closed against the orifice, pressure, applied to the larger area on the upper end of the spool, holds the spool

With fast response and little variation in repetitive consistency, this type of sensor is widely used where quality control, precise limits and positive physical proof of the presence or otherwise of the item needing to be sensed are important. It is also the type used extensively where fast cycling is needed—contributing to faster application and exhausting of signals than is normally associated with the use of internal, mechanically operated equivalents described earlier.

The fast response and positive changeover when initiated by a slow moving cam makes this type ideally suited for such work.

When designing cams to operate this type, there is nothing to be gained by depressing the mechanical device more than is necessary. Over-depression usually causes needless maintenance problems. Manufacturers' recommendations in respect to the required initiating movement should be followed closely.

Sensors responding to pressure change

The concept of a differential valve has been developed and expanded into a considerable variety of remarkably sensitive valves capable of responding to pressures as low as 0.3 kPa. Many of these employ a small spool or poppet arrangement with a constant pressure of normal main supply, say 550 kPa, applied continuously to a small area, while pilot pressure appropriate to the particular type— 30 kPa to 0.3 kPa—when applied to the pilot port of the sensor will change over the sensor mechanism causing the sensor to deliver a full strength, main supply, pressure signal. The pilot pressure is applied to a much larger area, often a flexible diaphragm, and the force resulting from the low pressure applied to the large area, being greater than the force resulting from the higher pressure on the smaller area, brings about the changeover of the internal mechanism of the valve.

Figure 7.7 shows the general arrangement popular with several manufacturers in achieving sensors which are sensitive to relatively

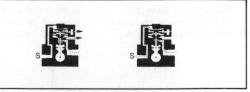

Fig. 7.7 *Diaphragm operated sensor, low pressure pilot air*

low pressures. Pilot pressures in this type will range from 5 kPa to 20 kPa. As may be seen, the pilot pressure is applied to a flexible diaphragm which closes a small orifice through which air from the main air supply exhausts to atmosphere when the orifice is open. Closure of this orifice concentrates the main supply pressure on a larger diaphragm which bears down on a differential spool or poppet piston. The smaller end of the spool is exposed to main air supply pressure, the force of which is overridden by the force derived from the pressure exerted on the diaphragm and larger end of the spool. Depression of the spool closes off the main exhaust port from the main outlet port, and opens the main air supply to the main outlet port.

Such devices, in which the pilot signal is of a lower pressure than the signal delivered finally by the device, are often termed either "step-up relays" or "amplifiers".

More sensitive devices of this nature are produced by bringing pressure to bear on the main operating diaphragm in two stages. In that shown in Figure 7.7, a small bleed, or exhaust, orifice is closed to bring full pressure to bear on a second diaphragm which causes the opening of main supply air to the main outlet port. The more sensitive types use this method of a small diaphragm closing a small bleed hole so that pressure is applied in full to a second such diaphragm, which closes a second port and causes the application of full pressure to a third diaphragm, which opens up the main supply to the outlet port. Pilot pressures required by such types as these can be as low as 0.1 kPa.

Another common type of low-pressure sensor is that shown in Figure 7.8. As may be

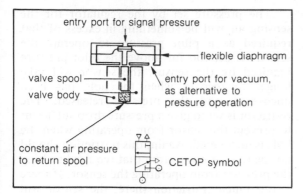

Fig. 7.8 *Diaphragm operation directly to spool, relatively low pressure pilot air*

Response times

As a general rule, it will be found that the more sensitive the sensor, the longer the response time will be. That is to say, although response to effect an "on" position may be very fast, reversion to the "off" position after the pilot pressure has been removed can be slow. Low-pressure air, in the pilot pressure areas of these sensors, flows at a rate proportional to the pressure. Thus, to exhaust a low-pressure pilot signal from such a sensor to its changeover point, when it reverts to the "off" position, takes longer than would be the case if the changeover point were higher in the pressure scale.

Most manufacturers quote response times in their catalogues and these must be taken into account when arranging the cycling time for a particular project.

seen, the pilot pressure is applied to a diaphragm which bears directly on a spool or poppet piston. This type usually requires pilot pressure of something in the order of 25–35 kPa.

Another type again is that shown in Figure 7.9 where pilot pressure is brought to bear on a small piston head. As may be seen in the sketch, pressure applied to the pilot port causes a piston to move the end of a lever. This causes the opening of a small orifice by the removal of a seal at the other end of the lever. Opening of the orifice allows pressure to be relieved to atmosphere from the larger end of a differential poppet spool. This movement of the spool closes off the sensor's exhaust port and opens up the main supply port to the outlet port.

The internal pressure-operated mechanism is very similar in its basic design to the external mechanical/internal pressure-operated sensor described earlier in Figure 7.6. Pilot pressures required by this type will be in the vicinity of 7–10 kPa.

Sensing the end of a cylinder stroke by the decay of the exhaust

Low-pressure sensors are frequently used for the purpose of sensing the end of a cylinder stroke. Assuming that, when all the exhaust in front of a moving cylinder piston has gone, the cylinder has physically reached the end of its stroke without suffering any hindrance en route, sensors such as those shown in Figure 7.10 are used.

Fig. 7.10 *Low pressure sensor, sensing decaying exhaust at end of cylinder movement*

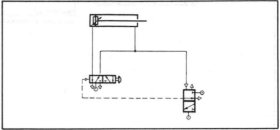

Fig. 7.9 *Air/bleed operated sensor, low pressure pilot air*

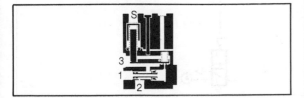

Used in this way, the sensor would not need to be one of the most sensitive types. In fact, since cylinder exhaust pressure should be approximately one-third of the driving pressure, the pilot pressure required by the sensor should not be further than a reasonable margin from the exhaust pressure so that the response time is reduced to a minimum.

For such applications, the type shown in Figure 7.8 would be preferable to the more sensitive types shown in other illustrations. Note in this case that the signal is derived when the pressure has been exhausted from the diaphragm rather than applied.

Jet sensing

All of these sensors requiring low-pressure pilot application lend themselves to "jet sensing". This term has been adopted to cover applications where the presence or otherwise of some object may be detected by its interference with a low-pressure jet of air. This interference may take the form of:

1. blocking or occluding a hole through which the jet passes, or
2. interrupting a jet blowing across a gap.

Figure 7.11 shows the general arrangement normally employed in setting up a sensor which will react to the blocking off of an air jet. The size of the hole emitting the air jet is normally about 3 mm diameter.

The pressure set by the regulator for the sensing air will be something in excess of that required as a pilot pressure to operate the sensor. If set too far above the pilot pressure required by the sensor, the sensor will take longer to recover when the hole has been uncovered and the pressure released. The restrictor is set to give a pressure drop sufficient to prevent the sensor from operating when the hole is uncovered. Again, this restriction should not be too far away from that required to keep the pressure from operating the sensor. If there is too much restriction there, the sensor will take longer than necessary to react when the hole is covered.

Operation

With the hole uncovered, the regulated, restricted air flows across the operating pilot port of the sensor and out through the hole to atmosphere.

When the hole is covered by the arrival of the device to be sensed, air from the regulator, with no means of escape, builds up pressure to the level set on the regulator. In doing so, the additional pressure is also applied to the pilot port of the sensor, which reacts to give a pressure signal.

When the device moves away from the hole, the pilot pressure is released and the sensor reverts to the "off" position.

Fig. 7.11 *Jet sensing–occluding low pressure jet*

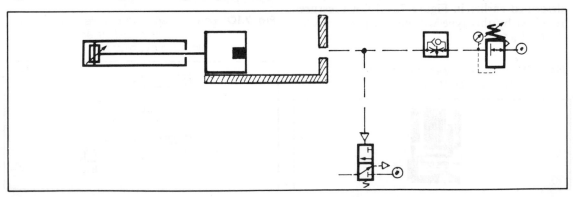

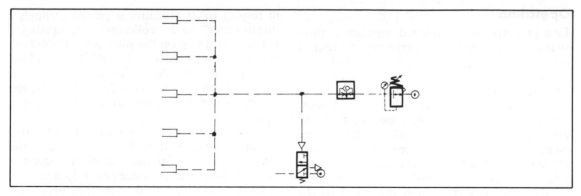

Fig. 7.12 *Jet sensing–occluding several low pressure jets to one sensor*

Practical pointers

Pipe resistance to compressed air flowing through from the sensor to the sensing hole plays an important part in the relationship of pipe length to sensor sensitivity. The more sensitive the sensor, the shorter must be the pipe. This will be due to the pipe resistance experienced by the air flowing through the sensing hole to atmosphere.

Adjustments of regulator and restrictor can usually be made at the final setting up of the project. Trial and error will indicate at what settings the maximum response, whether for "on" or "off", may be gained.

All pipes should be no longer than is necessary. Length, however, can vary according to the pilot pressure required by the sensor. For instance, a pilot pressure of 30 kPa can accept a pipe length maximum of upwards of 12 m, using 5 mm OD pipe. On the other hand, a sensor with a pilot operating pressure of 7 kPa will require a maximum pipe length, using 5 mm OD pipe, of 6 m.

The same general boundaries in respect to pipe lengths will apply where the one sensor is sensing the emission of air through a number of holes, all of which must be covered to operate the sensor (see Fig. 7.12). The use of fittings in distributing the sensing air to the various holes will further restrict the length of pipe from the sensor holes back to the sensor. All these factors will need to be taken into consideration. Usually, each application will need individual final adjustment on site. Thus, a margin for error and unforeseen contingencies should be allowed in respect both to pressures and pipe lengths. Some manufacturers can supply laboratory tested figures which will establish the boundaries within which to work, but the issue will be decided during the final on-site adjustments.

Proximity switch

One other method of sensing, which is akin to but not exactly the same as the method just discussed in which the air jet is blocked off entirely, is the method using a device known as a proximity switch. Mention of this type should be made before discussing the various methods of gap sensing. Figure 7.13 shows the basic design of a proximity sensor commonly used by a number of equipment manufacturers.

Fig. 7.13 *Pneumatic proximity switch sensor operation*

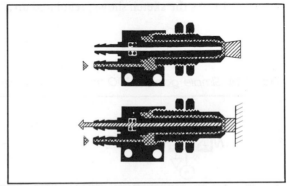

Operation

Low-pressure air is emitted through a ring-shaped orifice around the perimeter of a central orifice. The central orifice is connected by a small bore pipe, 4–5 mm OD, to a low-pressure sensor, amplifier or step-up relay.

On the approach to within 2 to 3 mm, depending on the type and pressures used, the emitted air is reflected back to the sensor, creating pressure in the central orifice and setting up a build-up of pressure back to the pilot port of the amplifier. This, in turn, changes the amplifier from the "off" to the "on" position and so a signal is delivered to establish the proximity of the object to be sensed.

The setting up, low-pressure sensor amplifier, restrictor and pressure regulator are as for the blocking of the jet escaping through a hole as in Figure 7.11. The proximity sensing device simply takes the place of the escape hole.

Gap sensing

The interruption of a jet blowing across a gap is a useful method of sensing where it is undesirable to make any sort of physical contact with the material whose presence or absence is to be established positively.

There are several different general arrangements of the main theme, namely transmission of an air jet across a gap to a receiver which is, in turn, connected to a sensitive amplifier or step-up relay. The signal from the amplifier is obtained when an object interrupts the jet by crossing through the gap.

The simplest arrangement is that shown in Figure 7.14. As the sketch shows, regulated air, in respect to its pressure, is passed through a "transmitter" to a "collector" jet, causing a pressure build-up on the pilot port of the low-pressure sensing valve. When the air jet across the gap is interrupted, the pressure at the pilot port of the sensing valve drops off, causing the sensing valve to change over and deliver a signal.

This simple type of arrangement is effective in gaps up to 100 mm. Beyond this limit the pressure required for the transmitter becomes impracticable. Performance above 50 mm gaps requires relatively high transmission pressures, so that the most effective results from this type of arrangement are usually found when transmitting across gaps below 50 mm. Such a gap, however, is amply great enough to cover a wide range of applications; for example, sensing conveyor belts when correcting any tendency of the belt to wander, sensing paper when winding it on to a spool, sensing venetian blind material in its processing, sensing the necks of bottles when labelling, capping, etc.

For gaps in excess of 50 mm or so, a gap sensing collector valve developed by Martonair Limited enables effective sensing to be carried out with the interruption of a jet transmitted across gaps as wide as 600 mm.

Figure 7.15 shows the operation of this collector valve. As the diagram indicates, the collector valve has an internal jet exposed to the gap by a relatively large orifice. The jet from the transmitter interrupts the jet within the collector valve so that there is no pressure from the collector valve to the pilot valve which will ultimately give the required signal. When the jet across the gap is interrupted, the internal jet in the collector valve effectively sends pressure

Fig. 7.14 *Simple gap (0–100 mm) sensing arrangement*

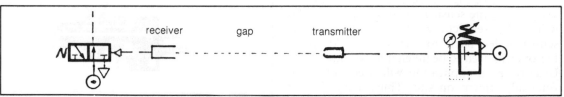

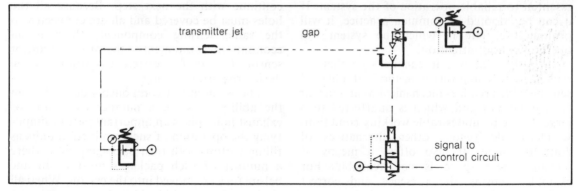

Fig. 7.15 *Sensing arrangement for sensing gaps up to 600 mm width*

through the connecting pipe to the pilot port of the low-pressure sensing valve to which it is connected. The sensing valve is thus changed over to the "on" position and a signal is delivered indicating the presence of the object to be sensed in the gap. Response times for this type are surprisingly good—50 milliseconds.

A third type is commonly used in dirty conditions where it is thought that the receiver jet may tend to become clogged with dirt from the working environment. This type is shown in Figure 7.16. As the diagram shows, low-pressure air, at a lower pressure than that provided for the transmitter jet, is emitted from the receiver jet. While the transmitter jet plays on the receiver jet, sufficient pressure is built up back through the connecting pipe to the sensor to hold the sensor in the "off" position. When

the jet across the gap is interrupted, the low-pressure outflow through the receiver jet is able to flow freely to atmosphere, resulting in a pressure drop at the pilot port of the sensing valve with a resulting change to the "on" position.

As with the arrangement first described in Figure 7.11, pipe lengths and pressures set by both regulators and restrictors are determined by the characteristics of the components used, combined with the final adjustments on site.

It is worth noting that whether or not working conditions are discernibly dirty or dusty, it is always good practice, wherever possible, to mount the transmitter jet facing upwards vertically to the receiver so that gravity will assist to keep contamination from entering the receiver jet. In dirty conditions this is

Fig. 7.16 *Jet sensing across gaps in dirty conditions, general arrangement*

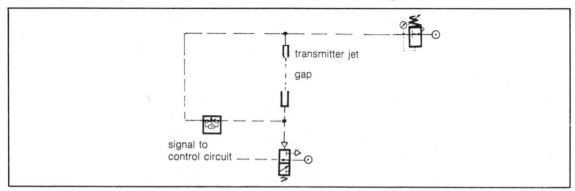

essential to reliable operation of the system. If it can be adopted as common practice, it will prolong the time in which the system will operate without attention.

From the above, it can be seen that jet sensing offers a positive method of physical sensing which reduces mechanical wear and tear to a minimum and which is unaffected to a large degree by undesirable working conditions which would render other alternatives of doubtful value. It also offers a means of sensing where space is not available. For example, sensing the presence and correct positioning of a small component in its processing can be achieved by a number of small holes which, when covered, provide the evidence that all is in position and the processing can continue with the next step. Because all the holes must be covered and all are connected to the same sensing component, there is an economy of components—the one component sensing four or five different positions of the device requiring sensing.

The economy of components derived from the ability to sense a number of escape or exhaust holes plays an important part in simplifying the operation of such production units as filling cartons with tins, packages, etc., where a number of such packages must be in line before they are moved into the carton. When all are in position and blocking their respective jet holes, the sensor, the one sensor connected to all the holes, will give the signal for the next movement in the cycle.

Planning to a productive cycle time

The cycling time of a productive unit is usually based on a predetermined minimum. This is set as the lowest acceptable level of output consistent with justifying its place as an economic unit in the overall investment which the total enterprise represents.

The unit may function as one of several steps in the processing of raw material into a finished product. In such a case it will have an output calculated to match that of the other related steps. Where it is a non-synchronised, independent unit, its production rate will be calculated against the acceptable maximum cost of the finished product. In either case, it is always highly desirable to aim at a reasonable excess margin above the minimum. Where it is one of several production steps, means of speeding up the various steps are usually evolved from time to time. Where it is simply a production unit in its own right, any improvement in its output which can be devised will eventually call for a faster cycle of operation of the unit itself.

In a pneumatically operated unit it is often an easy matter to arrange for a considerably faster cycle of operation than the initial concept calls for, then slow it down to the speed required at the outset of the project. The extra built-in speed usually costs very little more, if any at all, if it is built in at the start.

A typical example is that of a pneumatically operated bulk butter packer in a dairy factory. The original machine specified was based on a production rate of 250 boxes, or cartons, of butter per hour. The machine was designed to accept 20 kg blocks of butter from an extruder, wrap the blocks in paper, insert them in cartons, then seal the cartons and push them on to a conveyor for loading into the freezer. The rate of 250 cartons per hour was set to match the butter output from the churns and extruder. It also matched the estimated handling capacity of the staff manning the machine.

Within the first month of installation it was found that the butter output could be speeded up to produce just under 300 blocks per hour.

The handling staff, with practice, improved their capacity to match the output of the churns. The machine itself was speeded up to match the new demand by adjusting the flow regulators controlling the individual speeds of each of the forty-odd movements in the total cycle of operation. A slight increase in the speed of each contributed an overall increase in output of 20 per cent. A few years later, a continuous butter-making process replaced the churns and the transfer of the cartons from the machine was automated. The machine was then required to handle some 360 cartons per hour. Again, no change in the machine was required other than adjustment of the flow regulators controlling each individual movement. The final output rate was some 44 per cent greater than the original concept. The extra cost for this built-in excess margin was barely 5 per cent of the initial minimum capital cost, being the difference in cost of several relay valves controlling cylinders, together with the appropriate flow regulators and slightly larger bore pipe. All other costs were identical one way or the other. It pays to think ahead a little in matters of this nature.

When planning towards a specific cycle time, it should be remembered that the factors affecting the speed of any individual item in a cycle of operation are many and varied.

Cylinder loading, air flow and pressure, friction losses in pipe and fittings, response times of components, Cv factors of components, pipe lengths and the manner of piping up (direction of flow through tees, for instance), use of manifolds and, for that matter, the condition of the air supply, all combine to influence the final cycling time attained. Some of these factors can be determined. For instance, Cv factors should be readily available from manufacturers of components. Intelligent use of this information will enable the circuit design to incorporate matching flow rates. Other factors may not be so readily available in precise terms. Accordingly, to forecast a cycling time precisely would be an almost impossible task since the information required to do this would be difficult to obtain in its entirety. Thus the exercise is largely empirical.

Nevertheless, using a combination of laboratory tested facts and close approximations of other factors which have stood the test of time in practical conditions, it is possible to arrive at a forecasted time within fine limits. Especially is this so if the suggestion made earlier is carried out, namely, arrange for an estimated faster cycle then slow it down to the required time.

The constant striving of many engineers in widely varied industries to attain maximum productivity has established what might be called rules of good common practice. Those that are concerned with pneumatic operation are examined in the ensuing pages.

Productive sequence of operation

The first step in planning a cycle of operation is the arranging of the sequence in its most time-saving form. Wherever possible, non-productive movement should take place while productive movement is occurring. The process, for instance, should never be held up waiting for a non-productive return movement, say, of an ejector device. Often it will be found that the process can be separated into several subcircuits all of which can work more or less simultaneously. In such cases, the subcircuits would synchronise with each other only at the start and finish, each working independently in respect to its own particular cycle. In such cases, also, it is desirable to arrange the work so that each subcircuit will take roughly the same time to complete its work as each of the others.

An example of such a process is the forming of drain pipes from clay, a relatively lengthy process. The total process involves the extruding of clay on to a former. When the formed pipe has attained its required length, the extruder stops and the pipe is cut. A pipe handler device then moves into position, clamps the newly extruded pipe in its grasp and moves it over to place it on the shelf of a multi-shelf trolley. As each shelf on the trolley is filled with pipes, it is moved upwards and its place taken by an empty shelf. When all the shelves on a trolley are filled, the trolley is moved towards a kiln and its place taken by an empty trolley.

This process is divided into a number of independent subcircuits. These are synchronised only insofar as the completion of a subcircuit acts as the "all clear" for the following subcircuit to start. The first subcircuit is the movement of the former up to the extruder, the starting of the extruder, the dropping of the former at extrusion rate until the pipe length is reached, then the cutting of the pipe and stopping of the extruder. As soon as the pipe-handling subcircuit has removed the pipe, the pipe former returns to the extruder and the first subcircuit starts again on the forming of the next pipe. Meanwhile, the second subcircuit carries on with the task of laying the first pipe on a shelf on the trolley. On completion of its task, it returns to a waiting position, ready to pick up the next pipe from the former. The waiting time is short since the operations controlled by each of these two subcircuits take much the same time. The trolley subcircuits—moving of shelves and moving of trolleys—take place, when necessary, at the same time as the two first subcircuits are working. If a shelf has been moved, the empty shelf is usually in place ready to accept pipes by the time the pipe handler has picked up the pipe and is about to make the final movement in its cycle, namely placing the pipe on the shelf. This final placing of the pipe on the shelf will take place provided that a sensor indicates that the shelf is back in position, ready to accept a pipe.

By splitting the process into steps, each of which takes more or less the same time to complete, the overall time of the process is reduced to the time taken for one step. After the emergence of the first completed product, the time cycle is only that of the longest time taken by an individual step in the process.

Once the sequence has been positively identified and written down, the individual movements must be classified as productive or non-productive, reviewed as to feasible speeds and time required to complete their strokes in each direction, then studied in regard to the feasible overlapping which can be achieved.

Visual aids in establishing final sequence

While some find it sufficient merely to write the sequence of events down in the same manner as for cascades, others find it helpful to draw a "flow path" in which all events are coded with a letter or number and set out in a line as they occur. Others again prefer to use a "sequence diagram" in which time intervals are drawn and the sequence of events superimposed. Whichever method is used is immaterial. The important point is that all events should be written or drawn in the sequence in which they occur so that no event will be overlooked and all will come under close scrutiny as to its anticipated operational time and its relationship to the whole cycle of all steps in the process.

Practical example

A practical example of a simple automatic machine will serve to illustrate the various steps taken in their consecutive order when establishing an estimated cycle time. Figure 8.1 shows in sketch form the various stages involved in a machine which is devised to drill and tap a hole in a workpiece.

The sequence starts with the feed cylinder A pushing a blank workpiece from the feed hopper to a position under the drill cylinder C. Clamp cylinder B extends to clamp the workpiece. Cylinder C then extends to drive the drill into the workpiece. When the workpiece is drilled to the required depth, cylinder C retracts, after which cylinder B retracts to unclamp. Cylinder A extends to push a second workpiece under the drill. The first workpiece already under the drill is pushed by the second workpiece under the tapper. With both workpieces in position, both clamping cylinders, B and D, extend to clamp the workpiece. The drill cylinder C and the tapping cylinder E extend to preset strokes then retract. When both C and E are retracted, B and D retract to unclamp. Feed cylinder A then extends to push a third workpiece into the first position under the drill. The second workpiece is pushed by the first into its

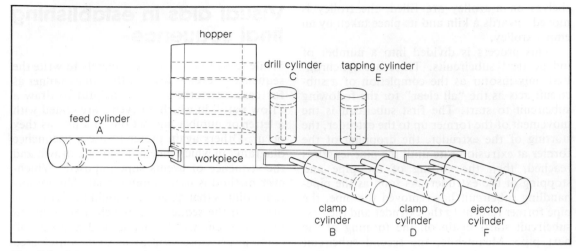

Fig. 8.1 *Practical example of simple automatic sequence*

new position under the tapping device and the first is pushed out to a position whence it may be pushed from the machine on to a conveyor belt. Cylinder *F* then extends to eject the first workpiece. Having done so, it then retracts.

The steps taken in arriving at a final sequence of operation follow very much along the pattern described.

Write the movements

Write all the movements involved in the total process, giving each a letter. The first movement is allotted the letter *A*, the second *B*, and so on. Write them down in the order in which they occur with a plus sign indicating an extension, if a movement, or the turning on of a valve, etc., and a minus if retraction of a movement or turning off of a valve.

In this example, Figure 8.1, writing down the steps produces in the first instance:

$$A + B + C + C - B - A -$$

A little thought at this stage brings up two points in the sequence as written down. There is no indication of what action is taken to place the workpiece in position to be pushed by the feed cylinder *A*. Furthermore, if the workpiece is to fall into position by gravity, what sort of

time will this take? The second point is in reference to the retraction of *A*. Is it necessary to wait for *A* to retract before another cycle can start?

Allotting the letters *W/P* to designate the workpiece, it must be written into the sequence.

To dispose of the time wasted waiting for *A* to retract and *W/P* to fall into place, the sequence can be rearranged. Since both these actions should take no longer than the time taken by *C* to extend and retract followed by the retraction of *B*, the sequence will now be written as:

$$A + B + C + C - B -$$
$$A - W/P +$$

Sensors combining to signal that *W/P* is in place and *B* retracted would provide the source of the signal for the cycle to start again, since, if *W/P* is in place *A* must have retracted.

When the second workpiece has been introduced, the sequence includes the actions of cylinders *D* (clamp) and *E* (tapper). These two cylinders perform their functions in much the same time as cylinders *B* and *C*. Thus the sequence becomes:

$$A + B + C + C - B -$$
$$D + E + E - D -$$
$$A - W/P +$$

Finally, in arranging the sequence, the introduction of the third workpiece brings in the operation of the ejector cylinder *F*. Since the ejection can be carried out any time after *A* has pushed in the third workpiece, with the simultaneous pushing into position of a workpiece under the tapper and into ejection positions, the sequence may be written thus:

$$A + B + C + C - B -$$
$$D + E + E - D -$$
$$A - W/P +$$
$$F + \qquad F -$$

The sequence would appear to be arranged so that there is the minimum of waste time.

Fig. 8.2 *Visual aid in sequence analysis*

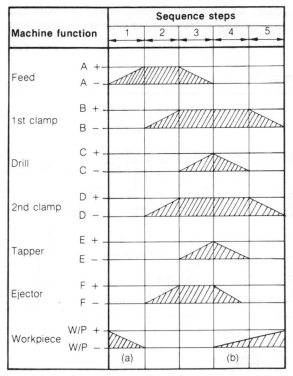

(a) workpiece fed into position by A

(b) Time available for workpieces to fall from hopper

However, a further saving can be made in respect to the two movements, retraction of *C* and *B* and the corresponding simultaneous movements, retraction of *E* and *D*. In these instances there is no need to wait until any of them have fully retracted. In the case of *B* and *D*, unclamping, the signal for the start of the next cycle, *A* +, can be given as soon as both have travelled far enough in their unclamping movements to release the workpieces. Thus the sensors can be placed accordingly. In the case of the drill and tapper, unclamping may take place as soon as both tools are clear of the workpieces. Again this is a matter of positioning the sensors to reduce waiting time here to the minimum.

At this stage some find it useful to draw up a visual check on the proposed cycle and its overlapping. This will help to indicate any further overlapping which may be possible. Figure 8.2 shows a method commonly used. The cycle is drawn up into approximate time units—say, of one second each.

Draw the control circuit

When a sequence of operation which will give the shortest possible time in the cycle has been obtained, the circuit should be drawn so that the time required to change over the various valves to operate the sequence may be taken into account.

In this example, a decision has to be made as to whether or not anything is to be gained by drawing a circuit which will cause the machine, when first started, to perform the first three movements, *A*, *B* and *C*, then return to the start position, and on its second cycle perform the movements *A*, *B*, *C*, *D*, and *E*, then return to the start position to cycle again. On this, its third cycle from start, it will perform all the movements of a complete cycle—*A*, *B*, *C*, *D*, *E* and *F*—and will continue to do so until it is stopped.

With a circuit designed to produce sequences for the first three cycles at start-up as just described, only those movements required to treat a workpiece would operate. The first cycle would take the first workpiece through its first stage of treatment. The second cycle would treat the first workpiece at its second treatment stage, while a second workpiece would be treated at the first stage. The third cycle would treat the first workpiece to its final stage—ejection—while the second workpiece would be at stage two and a third workpiece would receive its stage one treatment.

Alternatively a simpler circuit could be designed, provided no apparent harm would be done if the movements in the second and third stages could take place without a workpiece. With a circuit in which all movements worked in the sequence finally decided upon earlier, all movements would be dealing with a workpiece by the third cycle of the machine from start-up. Usually, if it is feasible to do so, the simpler circuit would be preferable.

Figure 8.3 shows a circuit which will operate the machine in compliance with the requirements originally described. As can be seen, the circuit lends itself to a simple two-group cascade system:

$$
\begin{array}{ccc}
\mathrm{I} & \mathrm{II} & \mathrm{I} \\
A\,+ & B\,+\,C\,+ & C\,-\,B\,- \\
 & D\,+\,E\,+ & E\,-\,D\,- \\
 & A\,- & W/P\,+ \\
F\,+ & & F\,-
\end{array}
$$

List time requirements

When listed with their times, the individual items which contribute to the total cycle time will be found to have their own particular characteristics. These characteristics are influenced by a combination of variables in each case. Consequently, each item will have an appreciable difference between the longest and

Fig. 8.3 *Circuit diagram for drilling and tapping machine*

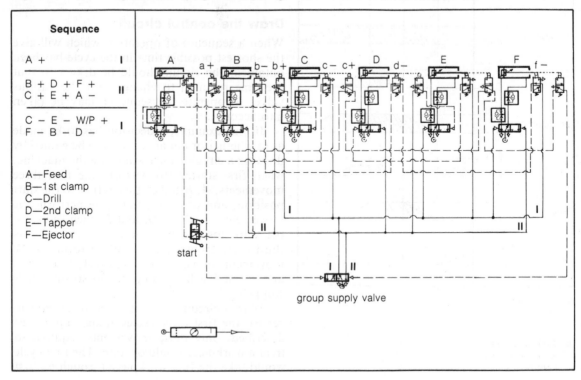

shortest time it may require to fulfil its task in the sequence of events of the operational cycle.

Before accurate estimates of these individual times can be made, the variables which influence them must be reviewed in detail. The sum of the individual items can only be reasonably accurate if the individual items themselves are so.

A detailed examination of the likely items which must be taken into account will enable the completion of the practical example shown in Figure 8.1 to be made with both understanding and accuracy.

Factors to be assessed in a cycling time estimate

Reference to the circuit shown in Figure 8.3 will show the individual items which must be taken into account in the sequence of events. When the start switch is set to the "on" position, a pressure signal has to fill a connecting pipe between the start switch and the relay valve controlling cylinder A. When the signal pressure is adequate, A's relay will be induced to change over. Air must then fill the pipe connected to A's rear end-piece, build up pressure on A's piston and drive that piston forward to the end of A's outstroke. When the sensor, $a +$, is depressed, air pressure from $a +$ must fill the connecting pipe to the pilot supply valve. When adequate pressure is in this pipe, the pilot valve must change over. Pressure must then build up in all lines connected to, and including, pilot line II. Relay valves controlling cylinders B, D and F must then change over. And so on ... to the end of the cycle.

Obviously, a time requirement must be known or an accurate estimate made for the time required for:

1. the build-up of adequate signal pressure in the lines;
2. the various valves to change over when the pressure is adequate;
3. each cylinder to complete its stroke;
4. the workpiece to fall into place from the hopper.

Factors affecting time requirements

The time required for the cylinders to complete their strokes will be governed by the feasible speeds at which they may move in relation to both the actual job they are required to do and the normal feasible controllable speeds of an air cylinder.

The work they are required to perform can be a limiting factor to be considered when reviewing feasible cylinder speeds. This work can be: drilling through various types of wood, hard or soft; drilling through metals, hard or soft; transfer of workpieces, loose, fixed, heavy, light; opening of hopper gates and closing of these, controlled or as fast as possible. It is always wise to allow a fairly wide margin in respect to the final feasible high level of speeds which may be attained after the plant has been in use for a time. By the same token, provision must be made for operating a little more slowly than may be considered necessary in the original concept. When taking the time estimate for the cycling time, a conservative estimate is wise.

The other limiting factor in regard to cylinder speeds, namely the actual feasible controllable speed of the cylinder itself, will depend on the type of cylinder used. Some standard cylinders have a maximum feasible controllable speed of approximately 1.9 metres per second. Others have controllable, cushioned speeds of up to 4 metres per second. Designing to a specific range of speeds is a simple matter of valve and pipe sizing, as described in *Practical Pneumatics*, Chapter 2. The time required to change over a valve is usually given in the manufacturer's catalogue. As a rule, it is included in the valve specifications under the term "valve response". Normally it is quoted in so many milliseconds.

These two items in the time schedule are relatively straightforward, i.e. cylinder speed and valve response. The question of signal time allowance is not quite so straightforward. It can be, however, an important factor in achieving maximum productivity from a manufacturing unit.

In the first place, consider the task of a logic component, or pilot component, in a control system. It must fill a length of pipe, between its outlet port and the pressure port of the valve receiving the signal, with air at sufficient pressure to change over the valve at which the signal is directed.

When the pilot valve opens to send the signals, the first of the pressurised air entering the pipeline expands very rapidly throughout the whole pipeline. As the flow of pressurised air continues to flow into the line, the pressure there builds up until it reaches the pressure at which the pilot air supply has been set. During this build-up, a point is reached at which there is sufficient pressure to move the mechanism of the valve at which the signal is directed.

So far as the valve on the receiving end of the line is concerned, the signal it receives will not be a full signal suddenly directed at it, but rather a build-up from atmospheric pressure to its required changeover pressure over a period.

What sort of time must elapse before the changeover pressure is received depends on five main factors:

1. the length of the pilot line;
2. the internal bore, or diameter, of the pilot line;
3. the Cv factor of the pilot valve;
4. the pressure of the pilot supply air;
5. the pressure required to change over the mechanism of the valve receiving the signal, coupled with its response time.

As each of these factors is dealt with in detail, the interrelationship of them all becomes very clear. Each needs equal consideration to achieve a balance by which all combine to produce the optimum in overall productive performance.

The length of pilot line

Every pilot valve has its own fixed flow rate. The greater the volume of air which has to pass through the pilot to fill the signal line with air at sufficient pressure to change over the valve at which the signal is directed, the longer will be the time taken to deliver an effective signal.

Obviously, the longer the line, the greater the volume required from the pilot valve.

Any pilot line which is longer than necessary is reducing the potential output of the production unit to which it is connected.

A high-speed production unit incorporating twenty consecutive steps in its process can be quoted as an actual practical example in which the output was increased by over ten per cent by cutting 0.5 m off every signal line. Signal lines are often left with significant length in excess of the minimum required for the job. It is important to remember that unnecessarily long signal lines will restrict the output of a unit for the remainder of its working life. Not only is the output seriously affected, but lines longer than necessary will also cause additional air consumption and, quite often, lubrication problems.

Internal bore or diameter of the pilot line

As with length, the bore size of the pilot line directly affects the volume of air which the pilot valve must deliver to provide an effective signal. It thus contributes directly to the time taken up by the signal both in its delivery period and in its exhausting period. This latter is equally important since the valve receiving the signal cannot react to another signal until the first signal is exhausted to atmosphere.

Occasionally, pipe of a bore size larger than necessary may be used under the mistaken impression that a smaller bore size will lead to harmful flow restriction and pressure drop. The flow, however, of a pressure signal is in a different category from the continuing flow to an air motor or to a moving air cylinder. The signal flow is more akin to the filling of a reservoir or container. The volume of air is flowing to a dead end. Thus, the bore size should be kept to a minimum so that the subsequent smaller volume in the line cuts down the signal time. Normally, no good purpose is served by exceeding an internal bore size of 4 mm diameter. Often a 3 mm internal diameter is sufficient. The deciding factor will be the Cv factor of the pilot valve employed.

Cv factor of the pilot valve

The flow rate of the pilot valve should achieve a balance with the bore size of the pilot lines. There is no useful purpose in using a pilot valve with a greater flow rate than the pilot line itself can accept. Conversely, where the flow rate of the pilot valve is less than the maximum flow rate which the pilot line can accept, the time taken to provide an effective signal is longer than would be the case where flow rates were evenly matched.

In practice, many of the so-called "moving part logic" components offered by manufacturers have Cv factors ranging from 0.14 to 0.2. With these a good practical balance can be achieved using 3 mm internal diameter lines. Where the Cv factor is down to something in the order of 0.026 to 0.06, smaller bore still is preferable if available, say, 3 mm outside diameter pipe.

Pilot air supply pressure

Naturally, the pilot supply pressure must be at least greater than the minimum pressure required to change over any of the valves in the system. Most main air supply pressures fluctuate between a compressor cut-in pressure of 550 kPa and cut-out pressure of 700 kPa. Frequently, it is found that the maximum changeover pressure required by any of the valves in the system is no more than 200 kPa. In cases such as these, i.e. maximum valve changeover pressure of 200 kPa and minimum main line pressure of 550 kPa, there is nothing to be gained by reducing the pilot supply pressure from that of the main supply of 550 kPa to a pressure closer to the valve changeover 200 kPa, say, to 250 kPa. On the contrary, such a reduction will add considerably to the signal time.

The greater the difference between supply pressure and valve changeover pressure, the shorter the time taken to change over the valves in the system. The reasons for this are twofold:

1. Where the pilot supply pressure is 550 kPa and the valve changeover pressure required is 200 kPa, air released from the pilot valve into the pilot connecting line at 550 kPa expands immediately throughout the whole length of the pilot line. The changeover pressure of 200 kPa will be attained throughout that line when only 0.46 of the actual volume of compressed air needed to fill the line completely with air at the full pressure of 550 kPa has passed through the pilot valve. The expansion of the air released enables the changeover pressure to be attained in just under half the time taken if the pilot supply air had already been reduced in pressure with proportionately greater volume.

2. The flow rate of air released from the pilot valve at 550 kPa into the pilot line will be sonic until the changeover pressure of 200 kPa is attained or surpassed. The difference between the two pressures is more than sufficient to maintain sonic flow until after the signal pressure has passed the changeover pressure required by the valve at which the signal is directed.

A useful formula to establish the differential minimum to maintain sonic flow through the orifice of the valve is:

When $P_1 > \dfrac{P_2}{0.528}$ sonic flow is achieved.

where P_1 = upstream pressure (absolute)
$\quad\quad\ P_2$ = downstream pressure (absolute)
and conversely:

when $P_1 < \dfrac{P_2}{0.528}$ flow is subsonic.

This matter of sonic and subsonic flow is important both in respect to the delivery of the signal—filling the signal line to the changeover pressure required by the recipient valve—and also the disposal, or exhausting, of the signal when it is no longer required.

A double pressure-operated valve which has been changed over by a pressure signal cannot be reversed by a pressure signal on its opposite pressure end until the first signal has virtually been exhausted to atmosphere. There can be no movement of the valve mechanism until the differential between the pressures of the two signals is in excess of the pilot pressure required to operate the valve. If the flow rate of the signal pressure exhausting to atmosphere through the orifice of the exhaust port of the pilot valve can be maintained at sonic flow, the disposal of the unwanted signal will take only the minimum period of time.

With the two aspects of a pressure signal in mind, i.e. rising pressure on one side and falling pressure on the other until the required differential is attained, there is a limited range of pressure, above or below which either signal delivery or signal disposal time must be unduly prolonged. The formula quoted above provides a useful check as to the desirable minimum pressure for each. It should be remembered, however, that the discharge orifice is a complex component in which the coefficient may vary from 0.62 for a sharp-edged nozzle or orifice to 0.98 for a well-rounded one. This means that a margin should be added to the differential figure provided by the formula to cover this variable contingency.

The following practical examples illustrate the method employed to determine the desirable pilot pressure as related to the changeover pressures required by the valves in the control system so that sonic flow is maintained for both delivery and exhausting of signals. In both examples the changeover pressure required for the valves in the system is 200 kPa.

To establish the minimum pressure which will exhaust to atmosphere at sonic flow:

$$P_1 > \frac{P_2}{0.528}$$

where P_2 is atmosphere 100 kPa (abs)

$$\therefore \ P_1 > \frac{100}{0.528}$$
$$P_1 > 189.39 \text{ kPa (abs)}$$

Therefore,

let P_1 = 189 kPa (abs) + 10%
= 208 kPa (abs)
= 108 kPa gauge

To establish the minimum pressure required to maintain sonic flow while delivering a signal up to a changeover pressure of 200 kPa gauge pressure:

$$P_1 > \frac{P_2}{0.528}$$

where P_2 is 200 kPa (gauge) or 300 kPa (abs)

$$P_1 > \frac{300}{0.528}$$
$$P_1 > 568.18 \text{ kPa (abs)}$$

Therefore,

let P_1 = 568 kPa (abs) + 10%
= 624.8 kPa (abs)
= 525 kPa gauge pressure

From the above, when pilot pressure is that of main supply pressure, 550 kPa, both delivery and exhausting signals will achieve sonic flow to the point at which the differential between each will be in excess of the changeover pressure required by the valve, namely 200 kPa.

With the advent of components requiring operating pilot pressures of under 10 kPa, the advantages of such sensitivity must be weighed against the time required to exhaust a signal before the component can be used again. The slower subsonic flow rate to achieve complete disposal of the signal may add to the cycling time.

Changeover pressure of valve receiving signal and its response time

As will have been obvious from the matters raised in the previous paragraph, the changeover or pilot pressures of components vary within wide limits. The components selected for a project must be considered from all aspects, not the least important of these being their pilot operating pressure requirements.

A low pilot pressure to turn a valve on will result in a speedy turning on, but can also result in a slow turning off, through slow exhausting of the "on" signal.

Response times are usually quoted by manufacturers for all components. In selection of components it is important to make sure that the response time quoted refers to a complete "on/off" performance, and is repetitively consistent in practice.

From all the factors influencing signalling time in a cycle of operation which have been examined in the preceding paragraphs, it can be seen that if rule-of-thumb methods are to be employed in assessing a cycling time, they should be used with care, discretion and an awareness that any error should be on the conservative side. It does not pay to be too optimistic when making an assessment for a project where the cycling time is critical.

In practice, the following rule-of-thumb time allowances have been found to be useful and close to the mark.

Cylinder movement

Depending on loading and valve/pipe sizing as selected for the individual work, plus pipe lengths between relay valve and cylinder of up to 2 metres, allow from 0.3 metre per second to a maximum of 1 metre per second.

Cam-operated sensors

As a rule, these are operated while the cylinder movement is still taking place. Under such circumstances their operating time may be ignored. However, in the case of very short strokes of cushioned cylinders, combined with the operation of certain types of sensitive bleed-operated sensors at the end of a cushioned stroke, some allowance, say 15 milliseconds, should be included.

Relay valves

1. *Double pressure-operated, ⅛″ BSP ported.* On average, changeover time for these small valves, whether used for logic purposes or as control valves connected to small bore cylinders, will rarely exceed 10 milliseconds.
2. *Double pressure-operated, ¼″ BSP and larger.* These will be found to average out at 15 milliseconds.

Valves: Pressure-operated spring return

These usually have approximately twice the response time of their double pressure-operated counterparts. Accordingly, for small valves

allow 20 milliseconds and for larger valves 30 milliseconds.

Solenoid pilot valves

Response times vary quite considerably from brand to brand. A reasonable average figure to use will be 12 milliseconds.

Solenoid-operated relay valves

In such valves there are the combined change-over times of the solenoid pilot followed by the relay mechanism. Thus a reasonable average in practice is found to be 30 milliseconds.

Signal time per metre of pipe length

This will depend on the matching of pilot valve Cv factor to pipe bore size, combined with pilot line pressure as compared with changeover pressure requirements of the valves receiving the signals.

1. Pilot lines matched to Cv factors, used to pilot valves requiring changeover pressures of 200 kPa or less, in which pilot supply pressure is maintained at 500 kPa or more, will return a time allowance for the signals of 10 milliseconds per metre of signal line.
2. Pilot lines matched as above in respect to Cv and bore factors, but having a differential of less than the 300 kPa between pilot supply pressure and changeover pressure requirements, will require a longer time allowance for the signals. A reasonable figure will be 30 milliseconds per metre.

All of the figures suggested presuppose the use of good quality equipment. Poorly manufactured equipment will show response times and inconsistency in repetitive performance far removed from the figures given. For instance, while some types of solenoid-operated relay valves have consistently shown changeover times of 0.025 second with almost no measurable variation, others have been tested which showed a variation alone in response of 0.038 second. Quality equipment is essential in achieving productivity.

Having established a rule-of-thumb time allowance for the various events which go to make up a complete cycle of operation, the working example referred to earlier may be completed both in respect to the listing of the steps and allocating a specific time to each step.

Referring again to Figure 8.1, a realistic appreciation of the work done and how it is carried out must be gained. On examination, it is obvious that the time cycle will include:

1. the movement of the workpiece from the hopper by gravity. This is designed to take place as soon as feed cylinder A has retracted. Before A can extend again, the two tappers, C and E, must complete the tapping strokes and then their retraction strokes. Thus, the movement of the workpiece can be considered to take place while other movements are carried out.

2. In general terms, it would appear that none of the movements can be carried out, desirably, at high speed. In detail, $A+$, $C-$, $B-$, also $E-$, $D-$, $F+$ and $F-$ could all move at a normal average of 0.3 metre per second. $B+$ and $D+$ would also move at 0.3 metre per second. $C+$ and $E+$ would move more slowly while drilling and tapping. Both the drilling and tapping movements would take about the same time as each other.

3. Signal times. These would be taking, in this case, 0.010 second per metre with an average length of signal line of 3 metres.

4. Valve changeover times. Double pressure operated relay valves controlling cylinders A, B, C, D, being ¼″ BSP, will require 0.015 second each. The logic valves and the relay controlling cylinder F, all ⅛″ BSP, will require 0.010 second each.

Turning now to Figure 8.3, the circuit diagram, the actual sequence may be defined and times allotted for each step.

1. Consecutive movement times:

$A + 250$ mm $B + 50$ mm at 0.3 m/s total	1.000
$C + 120$ mm at drill speed of 70 mm/s	1.714
$B - $ (part way) 10 mm	
$C - 120$ mm at 0.3 m/s	0.433
Total movement time	3.147 seconds

2. Valve changeover times:
A relay $+$, B relay $+$ and $-$, C relay $+$ and $-$, i.e. total of five large relay changes at 0.015 s each — 0.075
Pilot supply small valve two changes at 0.010 s each — 0.020

Total valve changeover time 0.095 seconds

3. Signal times:
Signals to be considered are:
Start signal to initiate $A +$
Signal from $a +$ to group supply valve
Signal from group supply to $B +$
Signal from $b +$ to initiate $C +$
Signal from $c +$ to group supply valve
Signal from group supply valve to initiate $C -$
Signal from $c -$ to initiate $B -$
Signal from $f -$, $d -$, $b -$ to start switch

These provide a total of eight signals, each of which is a part of the cycle time. The last signal could take as much time as two of the others since it has a devious path and a somewhat longer distance.

Allowing 0.003 second per metre and an average length of signal line of 3 metres, the total signal allowance would be nine times 0.009 second, i.e. 0.081 second.

The three totals combined:

movement	3.147
valve changeover	0.095
signals	0.081
provide a cycle time of	3.323 seconds
add for contingencies 10%	0.332
cycle time estimate	3.655 seconds

Once the unit had settled down to day-to-day production, the cycling time could be reviewed. It would be more than likely that the majority of movements could be speeded up. Altering the adjustment of the flow regulators controlling the individual cylinder speeds would probably cut the cycle time by something over 0.5 second. An additional 15 to 20 per cent output could result with no cost increase other than the additional air consumption.

Chapter 9

Pneumatic factors influencing quality control and productivity

Quality control of the end product and productivity are two indivisible aspects of any manufacturing project. Both are essential and both are so closely interrelated that often the factors influencing them for better or for worse are common to both.

Although it can be said that control of quality and productivity simply depend on the application of good commonsense, the day-to-day pressures on industrial engineers tend to blunt the edge of their perception until practices known to be undesirable are allowed to creep into the plant operations almost unnoticed. Familiarity with the immediate landscape dulls the awareness of any slow change taking place. Acceptance as a temporary expedient of one small compromise with the rules of good common practice leads to the first compromise becoming permanently built into the system when temporary X is added to the hitherto temporary Y.

The pneumatic influences on quality and productivity are here summarised so that, from time to time, they may be scanned. Scanning these will help to stimulate and revive that keen awareness of them all which must be present in the approach to each of the small day-to-day problems in any plant operation.

Quality control of the end product

The main contributing factor in achieving quality control from a pneumatic point of view is the ability to maintain consistency of performance in all pneumatic operations. Combined with this, of course, must be the inbuilt ability to adjust performance within the desirable limits dictated by the process carried out by the equipment.

The variables which can upset the consistency of performance lie in two main areas:

1. the main air supply, and
2. the component design, the selection and application.

Examples of variables related to air supply

Example 1

A guillotine, in which the cutting blade was driven down by an air cylinder, made an excellent, clean cut in the majority of its operations. However, at irregular intervals the cut would be found to leave a ragged edge on the product. On this account 20 per cent of the products were discarded as substandard.

Initially the blame for the poor cutting was laid on the operator of the machine, the cutting edge of the guillotine blade, and/or the raw material and its suppliers. All shared varying degrees of opprobrium for a long period as the quality control officers of the plant cast about for the real reason for such a high percentage of rejects.

Eventually, it was discovered that at frequent, irregular intervals the supply pressure dropped to 440 kPa when the blade cylinder commenced its working stroke. The guillotine had been designed to operate on an air supply pressure of 550 kPa. Working at 440 kPa the cylinder was at 25 per cent below design thrust. Cost to that company of a fluctuating main supply air pressure was 20 per cent of the value of the factory output—an extremely large amount of cash.

Example 2

In a new automated hospital laundry, an elaborate ironing machine functioned in a barely satisfactory manner between frequent periods of 15 to 20 seconds during which the ironing rollers were completely ineffective. Reject linen from those periods was piled up to one side and fed through again at what was normally the end of the working day. Staff then were paid overtime rates for an average of three additional hours per day while they disposed of the reject linen accumulated earlier through the day.

Final investigation proved that, although the pressure in the receiver from which the air supply was drawn was usually at the 550 kPa required by the pneumatic equipment relying on the supply, pressure at the point of entry to the machine in question rarely rose above 480 kPa and frequently dropped to almost no pressure at all.

As can be imagined, all other pneumatic equipment had also been designed to operate satisfactorily at 550 kPa so that nothing relying on air for its operation in that laundry worked effectively and often not at all.

Before the fluctuating pressure was remedied, cost to that hospital for those in the ironing section alone working unnecessary overtime periods amounted to a recorded 2340 working hours. Overall unnecessary costs not only included those particular figures but also many other hours on the part of engineers and other staff applying makeshift expedients in their endeavours to overcome the problem, together with expensive materials applied to no purpose.

Example 3

A concrete batching plant, with all its contracts requiring certified (to tight specifications) concrete was unable to maintain consistency of weight of the individual ingredients in the batches weighed into the mixer. All hopper gates were operated by air cylinders. Closing of the gates in response to signals from the scales was sluggish and erratic.

Investigation proved that, although pressure in the main receiver remained reasonably close to the desired 550 kPa, as each cylinder started to move the air supplied to it at the entry point to its relay valve suffered a pressure drop to a figure well below that for which the operation of the plant had been designed.

Loss to this company was a month's total production while remedial action was applied, after the cause of the trouble had been identified, together with an incalculable loss of customer goodwill in a competitive market.

Example 4

An air-operated bottle-filling machine, installed to cope with a seasonal high demand, behaved erratically. Partially filled and over-filled bottles were fed from the machine haphazardly while, at times, the machine would stop completely.

Investigation by an engineer, called in as a last resort, proved that air pressure at the point of entry varied from 550 kPa to 50 kPa. The machine had been designed for a consistent operating pressure of 550 kPa.

Loss to this company, until remedial action was applied, included the cost of a recorded 540 operator hours at overtime rates, plus unknown engineers' time and materials applying ineffective expediencies based on wrong diagnosis, plus incalculable loss of customer goodwill and unfulfilled orders in a highly competitive and seasonal market.

Example 5

A butter factory producing many tonnes of butter per day, with largely pneumatically automated equipment, suffered numerous unscheduled stoppages of the butter-packing section. Often, until the packer was brought back into operation, the butter was packed manually by the factory staff, working late into the night. The cause of the stoppages was not identified for some months. By that time overtime rates had been paid to almost the entire factory staff, averaging three hours per day for seven days per week. Labour costs for that factory over the dairying season almost doubled the normal amount per annum.

The cause was identified as contamination of the air supply by compressor oil carried over into the system from a small auxiliary compressor used for short periods each day.

Analysis of specific causes of variables

Example 1

Main supply source was inadequate for the consumption of the factory. Compressor, receiver, after-cooler and air distribution lines were undersized.

Example 2

Compresssor, receiver, after-cooler and one small section of the air distribution lines were undersized.

Additionally, the compressor was wrongly sited in the boiler house. Heat there was excessive. The compressor intake was also there, compounding the heat problem. The steam plant was undersized so that all steam relief valves were blowing off almost continuously as steam pressure was kept to the maximum. One such relief valve was situated 400 mm from the compressor intake. Thus saturated air at high temperature entered the intake. The high intake temperature was beyond the limits which the after-cooler could handle. The final result was a mixture of water and air at low to nil pressure in the air lines rather than the required adequate supply of clean, dry air at 550 kPa.

Example 3

All supply components were adequately sized—compressor, receiver, after-cooler and distribution lines. The cause of the pressure drops experienced by the cylinders was the excessive pipe lengths from relay valves to cylinders, combined with inadequate sizing of the pipes between valves and cylinders and also valves back to the main air supply lines.

Example 4

Although the compressor was adequately sized, it was wrongly sited. Although an air-cooled type, it had no ventilation and the intake was not ducted away to a cool place. With no after-cooler, saturated air passed at high temperature through a grossly undersized receiver to the bottle filler. There, the large volumes of water condensed to clog all pipes, valves and cylinders. The very small receiver, as related to the compressor size, caused almost continuous cutting in and out of the compressor. Efficiency was thus reduced considerably and the heat problem compounded.

Example 5

As already indicated, the cause of the erratic behaviour of the pneumatic equipment, to the detriment of quality and productivity in that factory, was contamination of the air supply from a small compressor which was overloaded. The compressor consequently ran without the normal off-load periods, which resulted in the overheating responsible for the oil carryover.

All the examples quoted were cases of substandard quality in the final product which was directly related to the quality of the air supply.

Fluctuating pressure results in:

1. fluctuating degrees of thrust and speed of prime movers;
2. erratic strength of control signals in the control system;
3. erratic functioning and response of control valves;
4. erratic signals from time-delay circuits and components;
5. erratic cycling times;
6. loss of control of the process.

Contamination of the air supply from excessive moisture or dirt, or carryover of compressor oil, results in:

1. interference with pilot signals;
2. erratic performance of valves and prime movers;
3. rapid deterioration of seals, swelling, sticking, etc.;
4. blocking of jets and bleed holes resulting in false or no signals, to the overall confusion of a total cycle of operation and loss of control of the process.

If these problems are to be avoided a constant check on the supply as related to the air consumption is needed. If new equipment is contemplated, the air supply must be reviewed in its entirety. The following basic requirements must be considered:

1. Sizing and siting of the compressor. Sized on load, or average consumption, plus 50 per cent. Sited in a well-ventilated, clean, cool place.
2. After-coolers must be selected to match compressor capacity.
3. Receivers sized to cope with maximum coincidental draw-off in terms of volume. (Rule-of-thumb method frequently applied is to take delivered capacity per minute of compressor, divide that figure by two and take the answer as the figure for the actual volume of the receiver.)
4. Sizing of the distribution lines to cater for the maximum coincidental draw-off in terms of rates of flow demands. (For a detailed description and chart refer to *Practical Pneumatics*.)
5. Provision of adequate moisture traps at all low points.
6. Provision of pressure reducers on all applications to maintain constant operating pressures.

In all, the provision of a satisfactory main air supply is a matter of absolute simplicity. It is a matter of commonsense and simple arithmetic. The dividends from setting up such a supply correctly are proven over and over again by many industries the world over. But to maintain a satisfactory air supply in a growing industry requires an awareness of the important part it plays in the overall picture of quality control and productivity.

Examples of variables stemming from component design, selection and application

Apart from the air supply, there is a less discernible area in which lie variables affecting the quality control of the finished product. This area is that of the characteristics of the components originally selected in the design of the machine. Lack of awareness of the existence of these variables has involved some companies in substantial losses of profit through the production of substandard products before any correct diagnosis of the cause was made. Typical examples are quoted here, taken from actual case histories.

Example 1

Metal objects were drilled to critical depths by a drill moved through the workpiece by a hydro-pneumatically operated cylinder. The depth of drilling was controlled by the positioning of a roller-operated/spring-return three-port valve which signalled for the return of the cylinder when depressed at the predetermined outstroke limit. The slow movement of the cylinder resulted in the slow depression of the sensor by a slow moving cam. The depth of drilling proved inconsistent.

Example 2

A factory, manufacturing household appliances, installed a new machine to cut lengths of sheet metal which were later folded into panels for refrigerator cabinets. The panel length sheet was determined by a microswitch making contact with the end of the moving sheet steel. On receipt of a signal from the microswitch, an air cylinder drove the guillotine blade down to make a cut. Variations in the cut lengths were found to be up to 12 mm.

Example 3

A machine designed to fill bags with a powder material up to predetermined weights of 20 kg varied in the final weights by up to 1 kg per bag.

Example 4

A timing device, controlling the contact time of a hot plate with the workpiece, appeared to vary so much that the process was almost abandoned in the belief that no timing devices of sufficient accuracy were available to control the critical time demanded by the process. Just in time to save the pilot scheme from being pronounced a failure, the real cause of inconsistency was identified and a thriving new industry established.

Example 5

An air cylinder was required to oscillate at high speed, the cycling time, cushioning and thrust all having a bearing on the end product. When originally set in motion, all three aspects of the movement progressively deteriorated as it oscillated until the movement finally stopped.

Causes of variables in examples quoted

Example 1

The slow depression of a roller/spring-return sensor has been described in Chapter 7, dealing with selection of sensors. Where any opportunity for mechanical friction is allowed to extend over a period of time, inconsistencies will occur. This was a case of wrong selection of the sensor and was rectified by the use of a roller/bleed-operated sensor with a movement of 0.08 mm instead of the 7 mm required by the roller/spring return. The internal pressure/bleed operation of the roller/bleed sensor also had a response time variation of less than 0.0001 millisecond.

Example 2

The variation lay in the design of the solenoid-operated relay valve controlling the cylinder operating guillotine blade. The type of seals used were of the static type and the valve when tested showed a variation in response time of 0.038 millisecond. When related to the speed of sheet metal feed movement, the 12 mm were accounted for in the variation of response. The fault was remedied by fitting a solenoid-operated relay valve employing seals of the loose O-ring type, passing over ports made up of a great many small holes rather than ports with large holes exposed to the seals moving over them. Under further tests, these showed a variation in response times of 0.0001 millisecond occurring every tenth operation. The other nine operations were so consistent as to place any variations which may have been there beyond the measuring capacity of the measuring instrument.

Example 3

The cause of inconsistency in this case was another type of solenoid-operated relay valve controlling the cylinder which cut off the flow of material when signalled from the scale head. Although employing a seal design slightly different in shape from that quoted in the previous example, the seals were still of the type generally designated as static seals. Fitting a similar solenoid-operated relay valve as that quoted in the previous example corrected the inconsistency, bringing the variation in weights down to under 90 grams.

Example 4

The cause of inconsistent timing, initially attributed to poor quality timers, lay in the type of relay valve used to remove the hot plate from the workpiece. Its variation was a characteristic of the seal design employed, combined with the internal finish of the valve bore surface. Fitting another valve with a consistent response time remedied the condition.

Example 5

This condition was described in Chapter 7 and was due to the selection of the wrong type of sensor for the work involved. As in example 1, the sensor originally selected was a roller/spring-return type. The remedy applied was to fit a roller/bleed-operated sensor.

Many other examples could be quoted of incorrect selection of components for the work involved. Several cases of air-operated friction clutches and brakes can be quoted as performing inconsistently, until the valve controlling them had been sized to supply large volumes of air in short times, so that when pressure was applied there was little time for the friction to play before all movement had either ceased or come up to full speed, whichever the case might be. Either way, mechanical friction was the underlying cause of variations in performance and fast application or exhausting of pressure brought the time for friction to work down to the minimum possible time.

It should be clear that components all have different inherent variations in response times and operating characteristics. Study of these before final selection is imperative if quality control is to be built into the machine. Only

their consistency in functioning gives a machine the ability to process the raw material through to its final state as a quality product which can hold its own in a competitive market.

Productivity influenced by pneumatics

If we regard the term "productivity" as expressing the elimination of waste time in processing the raw material from its original state to the finished product, to secure maximum productivity from pneumatic equipment it will be necessary to examine a fairly wide area. Some of the points listed here may appear to be obvious but they can be overlooked by even the most competent in the stress of the moment. Accordingly the list is provided as a reminder, a check list or a shopping list to serve as a means of comparing an existing state of a plant against the desirable state to which everyone is striving.

Sequence of operation and control circuit design

The base concept of how the processing will be carried out is always the first to examine in a new project. It must not only be examined for feasibility but also in detail to ensure that every step in the sequence is fully productive. In other words, every step in the process must move the product forward in its processing towards the finished article. Any dead movement must take place while the productive steps are in progress. This has been discussed in detail in earlier chapters when dealing with circuit design. It must, however, not be left off the check list as it is at that stage that productivity can be permanently made or marred.

Selection of components

The selection of components must be made after consideration of many different aspects, as they are related to each other and to the first cost together with continuing cost. These aspects include:

1. functional effectiveness—this is of first priority.
2. repetitive consistency—a must for quality control and elimination of waste time through variable response.
3. availability—a great deal of valuable production can be lost waiting for something which never arrives.
4. availability of spares—sometimes better to install a slightly inferior component than risk long periods of down time waiting for spares to be delivered from an unreliable source.
5. life expectancy—obviously better buying to buy something at twice the price but three times the life of an alternative. Apart from initial cost, the cost of down time is a great deal less if the component has to be replaced only once every two years rather than three times in that period. One extra down time period would probably pay for the whole machine in many cases.

Layout

A great deal of time can be wasted if insufficient thought is given to the following.

Accessibility of all wearing parts

Those parts which require frequent replacement should be given especially careful consideration so that they can be replaced quickly and easily without having to remove other components.

Accessibility of all fittings

This requires careful thought, not only to ensure that pipes may be removed individually, when required, without having to remove others to do so, but also to ensure that a spanner may easily reach every fitting and every fitting may be easily tested for leaks.

Accessibility of all control components

It is important that replacements can be made quickly with the minimum of down time to cater for the regular maintenance these items will require in respect to seal replacements, adjustments, etc.

Ease of identification of power and control components

These must be easily identifiable so that they can be related to a circuit diagram by the maintenance staff with the minimum waste of time in the event of any stoppage, planned or unscheduled.

Eye appeal

It is generally recognised that it is false economy to dismiss eye appeal as of no importance in the financial welfare of the factory. While the immediate effects of ensuring that the machinery both functions well and looks well may appear to be largely psychological, these effects can be far-reaching in their final contribution to the economic robustness of the enterprise.

These psychological effects are twofold. The sight of a well-cared-for, neat and tidy factory, using well-finished, efficient-looking machinery in the manufacturing of a product, ensures both confidence in the product itself and enthusiasm on the part of those who may be considering investing in the marketing of the product. It implies that the manufacturers really know what they are about. It demonstrates that they are well equipped to maintain the quality of the product and take in their stride the demands for increased quantities which will arise from successful marketing.

Eye appeal has also an important influence on those who will be required to operate machinery and also those who will be responsible for the machine's day-to-day maintenance. An untidy shambles of a machine will gain scant respect from either operators or maintenance engineers. Operators will be reluctant to study seriously the ways of gaining the best results from the machine. The engineers will regard it as an insult to their intelligence, shunning any contact with it. Thus, even such essentials as the daily checking of filters and lubricators will be neglected. Because of this, the machine will have a short and troubled life, ending up as a white elephant in the factory museum.

On the other hand, if the machine both looks well and performs well, it becomes something in which both operators and engineers take pride. Such a machine will never lack for care and attention. It will continue to pay its way long after it has been depreciated to a nil figure on the company's books.

Maintenance equipment

Machines standing idle while some form of malfunctioning of pneumatic equipment is remedied may do so for unnecessarily long periods if the maintenance staff are not equipped with the facilities to carry out repairs quickly and effectively. Loss of production caused by such maintenance equipment deficiencies is not always clearly recognised. Pneumatic equipment does not normally require a great deal of specialised equipment and such as is required is rarely expensive. The following suggestions have proved their worth in practice in many industries.

Pneumatics workshop bench

Because dirt is the common cause of many small unscheduled stoppages, an awareness of the importance of cleanliness has to be cultivated and encouraged among the workshop staff until it becomes a habitual mental attitude for all concerned. A major aid towards this is the establishment of a special pneumatics work bench in the workshop.

Such a bench must be reserved solely for the repair and assembly of pneumatic components. It must be kept absolutely clean and free of such likely contaminants as swarf. Components, before they are placed on the bench, should be wiped clean. Before working on the bench, maintenance staff must develop the habit of washing their hands.

All of these precautions, if carried out meticulously, will ensure that no cause for stoppages in the form of dirt or swarf are introduced while the components are undergoing service.

Naturally, the bench will have appropriate air supplies for testing, together with gauges, pressure reducers and the like. It should also be equipped with the tools which make servicing of the components easy, such as:

1. wooden contoured blocks to hold valve spools in a vice while the operating mechanism is removed from the spool (sets of these should be held to cover the common sizes of spools used in the plant);
2. special tools for fitting seals to valves and cylinders;
3. circlip pliers to fit commonly used circlips.

Circuit diagrams

Circuit diagrams, correctly drawn and up-to-date, must be readily available for use in day-to-day fault diagnosis. These diagrams may be fixed to the individual systems they represent, perhaps on the inside of the control cabinet door, or they may be reduced to a convenient common size, glued to a suitable stiff material and covered with transparent plastic to keep them clean. They may then be filed in a cabinet in the workshop for day-to-day use in fault diagnosis. Every pneumatic operation in the plant must have such a diagram. The saving in lost production time through quick and easy fault diagnosis—only possible when a diagram is available—can amount to astronomical figures.

Maintenance staff training and knowledge

Adequately trained staff, obviously, are important in the productivity of a plant. However, the simplicity of pneumatics has often led to the misconception that little or no instruction is necessary for those looking after pneumatic equipment. Its simplicity cannot be denied but there are certain fundamentals which must be known by the maintenance personnel on the floor if they are to look after the equipment intelligently and effectively. The following are certain aspects of pneumatics with which all personnel must be familiar.

1. The nature of compressed air and the laws of nature which determine its behaviour when it is put to work.
2. The idiosyncrasies of the flow of compressed air through pipes and orifices at various pressures.

3. The relationship of pressure to load in the operation of a double-acting air cylinder. Staff should know this so well that they can recognise how and where this occurs in practice.
4. The control of thrust and speed in pneumatic prime movers of all kinds.
5. Significant design features and functions of commonly used components.

Maintenance staff must have the ability to read and interpret a circuit diagram. This implies a knowledge of CETOP symbols and the pattern of thought behind the making up of a circuit, together with cascade up to a three-group system. Staff should be trained in the practical use of a diagram in diagnosing the causes of unscheduled stoppages.

They should also have instilled into them a keen awareness of the importance of a clean dry air supply. Coupled with this must be the ability to recognise in practice the underlying causes of common stoppages due to contaminated air or faulty lubrication. Not only should they be able to identify the real, underlying cause of a stoppage, rather than a superficial cause, but they must recognise the importance of reporting the real cause of a stoppage to those in charge.

Armed with instruction on these points, the engineering maintenance staff will be well equipped to handle the day-to-day operations of the pneumatic power and controls in their care. Experience will complete the development of their expertise combined with intelligent thinking, enthusiasm and liking of their work.

So far as professional and technician level engineers are concerned, as well as familiarity with all of the aspects listed above, they will need to explore the depths of pneumatics a good deal further if they would derive the full potential pneumatics has to offer in attaining low cost automation and economical productivity.

One other aspect of the upper echelon engineers' obligations has been referred to earlier in this book, but it is of such importance that it cannot be left out of this list. This is the need for an appreciation of the necessity and the ability to spell out the economics of all work done. For instance, the loss of production caused by a substandard air supply—contaminated with water and compressor oil, and subject to serious fluctuations of pressure—needs to be transposed into dollars and cents if the funds required to rectify the situation are to be made available. Often the cost of remedying such problems appears to be considerable until it is related to the cost of production lost through leaving things as they are. Such a comparison will prove that the remedial costs are but a small percentage of the invisible but tremendous continuing drain on the net profits of the whole enterprise.

Communication

In every sphere of activity in this increasingly complex society of ours, communication, vertical and lateral, is more and more difficult to maintain. Yet communication, as much as any other aspect of an organisation, can spell success or total failure.

Within the engineering group in an enterprise, the lines of communication must be kept wide open—vertically in both directions and laterally at all levels—if the true picture of the plant's operations is to emerge and continue to be seen in total perspective by those who control its future in the final issue.

In dealing with pneumatics, there are some essential matters which depend on good lines of communication. Some of these are as follows:

1. Information about the detailed performance of the pneumatic equipment comes best from those who are in daily contact with it—the tradesman, the fitter, who attends the stoppages, dismantles the valves and so on. In larger plants where these personnel work on shift, it is usual for them to fill out a report at the end of each shift. In smaller plants any reports which may come through will usually be verbal.

   Whether these reports are verbal or written it is important that those who make them are made to feel that these reports play an essential part in keeping the plant running at peak efficiency. They must know that their reports are read, assessed and co-related. They must know that the information they provide assists in the determining of any action taken later. If they cannot be sure this is so, such a vital source of information can easily dry up.

2. In any report of a malfunction in pneumatic equipment, it is important that the true cause of the malfunction should be identified and conveyed. For example, it is not sufficient for a fitter in a shift report to state that the seals have been replaced on such and such a valve on machine X. Such a report is literally of no value unless "because . . ." is added. The seals may have become worn through the fair wear and tear of ten million operations. On the other hand, their life may have been much shorter through lack of, or poor, lubrication. Or again they may have become swollen long before their life expectancy through contamination from compressor oil. If these things are not stated, the reports do not represent a true picture of the plant's daily performance. In fact, the picture can be completely erroneous. Thus, the maintenance staff on the floor require training to recognise the symptoms and encouragement to report on their diagnoses.

3. In reporting stoppages and the reasons why, it is also essential that the length of time of each stoppage be reported. It is equally essential that such reports should be analysed and correlated. From this, accurate records and calculations of the value of loss of production can be made, to support any expenditure required to remedy any undesirable situation. Cases can be quoted of plants doubling their output without any major expenditure in capital equipment or increase of staff simply through accurate reporting followed up by remedial action. In one such particularly successful case, the chief engineer took his whole staff into his confidence, displayed

for all to see graphs showing plant stoppages through malfunctioning of any part of the equipment in their care together with time and value of production lost.

4. While in the main the reporting must come upwards, there is a real need for communication to travel downwards, especially when any modifications or improvements are contemplated. Those working closest to the machine will often have worthwhile ideas on the subject which, even if not so practical as they may imagine, often spark off ideas on another tack which are practical and worthwhile.

Interdisciplinary communication sometimes may be referred to as interdisciplinary respect. Whatever one may call it, it is a matter which can receive less than the attention required by engineers.

Whether a plant be large or small, any expenditure must be justified in relation to its contribution to the whole economic structure of the enterprise. Too often engineers become frustrated by having their requirements ignored through no other reason than that they have done insufficient homework on the particular problem. Each problem has to be approached both from the economic angle and the angle that no one can agree to any proposition unless he can understand it. Only a fool will allow himself to be "baffled by science" and few top executives are fools. The proposition has to be explained in terms which may be readily understood by the average layman.

Communication, knowledge, respect and an awareness of the interdependence of all upon each other in an organisation are the essential ingredients of a successful enterprise.

Pneumatics has much to offer in any industry. Knowledge and understanding at all levels are the keys which will open the doors to its full potential.

Chapter 10

A case history of a pneumatically operated process

The question may well be asked as to how the widely different aspects of pneumatic power and control may be combined in the solving of a practical problem, the answer to which appears to lie in pneumatics. The following case history is quoted in some detail, from the original concept to the final commissioning and maintenance stages. The various stages through which the project passed to final completion will serve to illustrate the blending of many different aspects into a pattern which desirably should be associated with any new project.

The problem described here was the devising of a machine to pack automatically blocks of butter, each weighing 20 kg, as they emerged from an extruder in their final marketable state. The machine had to wrap the blocks of butter in a special wrapping material, place the wrapped blocks in cartons, glue the carton

flaps and seal the joins with tape. The sealed cartons were then to be transferred to a conveyor which would load them into refrigerated rail or road wagons for transporting to the export refrigerated stores.

The machine which eventually successfully conformed to a performance specification laid down by the dairy products manufacturing industry is still manufactured and used internationally. While its appearance has improved, the basic design has required little or no modification. Its success, as compared with other attempts to produce an alternative machine, lies in the attention which was given to all stages of the project in its transition from a performance specification on paper to a production unit in an almost fully automated butter factory. These stages and their implications may be examined in the ensuing pages.

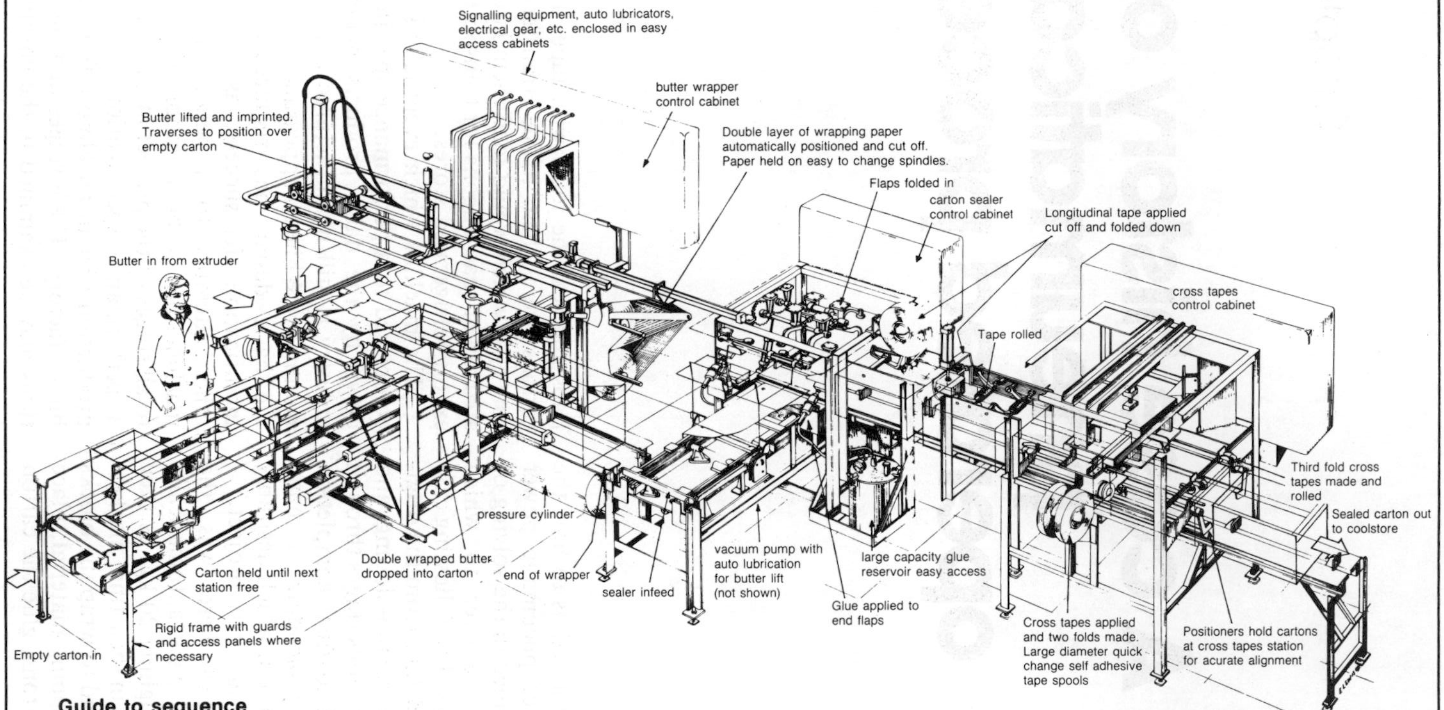

Fig. 10.1 *Butter packing process*

Signalling equipment, auto lubricators, electrical gear, etc. enclosed in easy access cabinets

butter wrapper control cabinet

cross tapes control cabinet

Third fold cross tapes made and rolled

Sealed carton out to coolstore

Positioners hold cartons at cross tapes station for acurate alignment

Longitudinal tape applied cut off and folded down

Tape rolled

Cross tapes applied and two folds made. Large diameter quick change self adhesive tape spools

carton sealer control cabinet

large capacity glue reservoir easy access

Glue applied to end flaps

Flaps folded in

Double layer of wrapping paper automatically positioned and cut off. Paper held on easy to change spindles

vacuum pump with auto lubrication for butter lift (not shown)

sealer infeed

end of wrapper

pressure cylinder

Double wrapped butter dropped into carton

Carton held until next station free

Butter lifted and imprinted. Traverses to position over empty carton

Rigid frame with guards and access panels where necessary

Butter block lifted and imprinted under butter. Butter traverses.

Butter in from extruder

Empty carton in

Guide to sequence

1. Formed carton with top flaps open, fed into machine by gravity or power conveyer. Machine accepts cartons when required. Carton positioned under butter.

2. Butter block lifted and imprinted by vacuum pad. Butter traverses.

3. Double layer of paper pulled into position under butter and cut off to length.

4. Butter placed on wrapping paper.

5. Wrapped butter drops into waiting carton.

6. Cartoned butter passes to sealer. Carton enters glue station where end flaps are glued.

7. End flaps followed by side flaps are folded in.

8. Longitudinal tape is applied and rolled on. Tape is cut and folded down.

9. Longitudinal tape is rolled.

10. Cross tapes are applied and cut off. Side and end folds made.

11. Third tape fold made and tapes rolled.

12. Sealed carton ejected.

As operations overlap the above sequence is true in principle only.

Characteristics of the raw materials to be processed

The success of the project required, in the first instance, a close knowledge and appreciation of butter as a material to be handled by mechanical means.

There is on record a prototype butter packer developed by a group which lacked this appreciation. The machine could wrap and pack in cartons blocks of wood shaped to simulate blocks of butter, beautifully. However, when butter was fed into the machine it seemed to lose all sense of coordination. After a brief period of hysteria, it invariably stopped. This particular group eventually abandoned an exceedingly expensive project—a project which failed simply through lack of real knowledge of the raw material.

The questions asked at this stage related to hardness, softness, tendency to slump, weight for bulk ratio, stickiness, moisture content, and so on. Together with all these were the variations in each which might take place throughout a working day, brought about by temperature changes, or by the changes which there may be in manufacturing a product of many slightly different types to cater for the differing tastes of a world market. Butter, like every other natural product, has its inherent idiosyncracies which can only be learned by a period of personal involvement.

In this case, from the intimate knowledge of these characteristics possessed by engineers experienced in the dairying industry, the overall mechanical design was evolved. For instance, the varying degree of hardness of the newly extruded butter was known to be such as to make its lifting by application of a vacuum pad feasible.

Once the mechanical concept had been established in general terms, a prototype was built to test the practical operation of the vacuum pad lifting device and the wrapping paper folding apparatus. These operations were tested under practical conditions in a dairy factory.

Selection of suitable control and power media

As the mechanical concept took definable form, a decision had to be made as to the means of power and control. To determine this, several interrelated aspects were given consideration.

Working environment

Under this heading came the question of the sorts of conditions under which the equipment would have to work reliably. A dairy factory is subject to stringent regulations in regard to hygiene and safety. Such regulations demand that all machinery must be washed down at frequent intervals with specified cleansing materials, including caustic soda, and copious volumes of steam and hot water.

Under the same heading consideration was given to the typical maintenance staff in such factories. Because understanding of the process is important for engineers looking after the manufacturing equipment in such establishments, the maintenance staff usually graduate through the ranks of the butter makers, thence into the engineering maintenance section. Since the work they do is largely of a mechanical nature, any control system should desirably be such as can be readily understood by the existing staff.

Production rates

These vary from factory to factory, so the power units, desirably, had to be capable of functioning within relatively wide limits of speed. Changes of speed needed to be controlled by simple adjustments able to be carried out on the spot by operators with only a lay approach.

Equipment

All equipment selected needed to be compact to allow overall design to conform to the space available within the average factory layout.

There had also to be a reasonable supplier back-up, in regard to technical knowledge and availability of spare parts.

Power units were required to be such as would lend themselves to fine control in respect to thrust and speed because of the nature of the product handled.

For instance, the vacuum pad, when applied to the top of a block of butter prior to lifting it, was required to apply sufficient pressure to imprint a brand design in relief on the surface of the butter. The thrust had to be controlled so that the imprint could be effectively made without causing the butter block to slump or butter to be forced into the vacuum suction orifices.

Discussion of all these aspects finally resulted in the decision to operate both the power units and the control system pneumatically throughout.

Once the decision to operate entirely pneumatically had been made, a pneumatics engineer was invited to join the project development. He was charged with the task of specifying the type of power units required, designing a suitable control circuit, advising on the selection of items of equipment such as sensors, which would be exposed to the less pleasant elements in the working environment, and also advising the means of protection required by the equipment, general layout and method of installation and connecting up of the pneumatic equipment employed. Included also in his responsibilities was the writing of operating and maintenance instructions for the pneumatic equipment section of the instruction booklet supplied with each machine.

Control design

The pneumatics engineer's first task was to acquaint himself thoroughly with the conception of the machine as a working unit which existed in the mind of the mechanical engineer responsible for the project—not as it existed in prototype form in its incomplete state.

This exercise developed into a two-way interchange of concepts—mechanical to pneumatic and pneumatic to mechanical. The mechanical engineer found himself reviewing his mechanical movements to ensure that they complied with the behaviour of an air cylinder. Alignment tolerances, guiding of movements, changes of load condition, speeds and thrusts all came under review by the two men who very quickly established a mental rapport so that each could claim to have as nearly identical images of the machine as is possible.

With the nature of the mechanical movements approved by both, the sequence of movement was discussed.

First, the mere practical requirements of the process were outlined, identified and written down. The steps in the butter packing process, from extruder to final conveyor are as follows:

As the butter leaves the extruder, it is cut into blocks and check weighed.

When the packer is ready to accept a new block, the block moves on to a reception platform.

When the block is in the correct position on the platform, a flap, depressed by the presence of the block, activates a sensor.

If the packer is ready, an air cylinder extends, pushing down to the block a vacuum pad which serves the dual purpose of printing a brand design on the block and acting as a lifting device for the block's transfer to its next position.

When the contact has been established between vacuum pad and block, and sufficient vacuum has formed, a sensor signals the vacuum pad cylinder to retract and lift the block vertically.

The next step in the wrapping process is the transfer of the block to a position above what is termed the "folding table". The folding table has a number of flaps, each held in position by an air cylinder. While wrapping is in progress, the table flaps are held to form a flat wrapping platform.

Before the block can be brought down to the wrapping table, the special wrapping paper must be spread over the table and held there until the block is placed on it.

When the paper is spread across the table—pulled from a roll of paper—it must be cut after mechanical "fingers" have secured their hold on the four corners of the spread paper. The sheet of wrapping paper is then ready to accept the block.

At this stage, paper cut and ready on top of the folding table, the block, suspended from the vacuum lifting pad, together with the lifting cylinder, is transferred by a traverse air cylinder to a position above the wrapping paper.

If the paper is not torn, the block is lowered on to the paper. Vacuum is then cut off from the lifting pad and low-pressure air blown through the vacuum orifices in the pad. This ensures that the block is freed quickly and completely from the lifting pad. The lifting pad is then raised and the pad and its lifting cylinder return to their original position above the receiving platform. With the block on the wrapping paper and the vacuum pad out of the way, the appropriate flaps on the folding table move in sequence, folding the wrapping paper neatly about the block, pressing the folds so that they remain in their correct folded state.

Before the next step can take place—the placing of the wrapped block in a carton—a carton must be prepared and placed in position under the folding table.

Once the carton is ready in its correct position and the lifting pad has moved out of the way, appropriate flaps in the folding table open up so that the wrapped block is guided, as it drops by gravity into the waiting carton.

The folding table then returns to the ready position while the carton containing the butter block is pushed by an air cylinder into the carton gluing and taping section.

As the carton is pushed into the gluing section, the flaps are held in position and glue applied to the appropriate areas. The flaps are then folded and pressed to ensure satisfactory adhesion. The cartons then pass on through the taping section where sealing tape is applied to the flap joints.

All movements in the gluing and taping areas are carried out by air cylinders operating in correct sequence, designed so that no movement can follow another until the preceding movement is complete.

As each carton emerges from the taping section, an air cylinder pushes it on to the conveyor which will load it into the transport.

Designing a sequential control circuit combined with maximum output

A study of the list of events which had to take place in the process clearly showed that there was no need to wait for each block to emerge wrapped, cartoned, glued and taped before another block could be processed. In fact, by dividing the sequence into a number of sub-sequences, virtually all non-productive movements were able to take place while other productive movements were advancing the process through further steps. For example, the forming of the cartons was treated as a separate circuit, interlocking with the packing, so that, so long as a carton was in its correct position by the time a butter block was wrapped, the folding table signals to open and drop the block into the carton would be delivered. While the wrapping takes place, the lifting apparatus is returning to take up its position over a new block arriving on the reception platform. Thus, three or four blocks are usually found in the machine together, each at a different stage of processing.

Having agreed upon a final sequence of events, satisfactory to all involved in the project, the sequence was written down in the manner described in the chapter dealing with the cascade system of sequential control— $A+$, $B+$, etc.

Emergency and fail-safe action

Questions of emergency and fail-safe actions were dealt with at considerable length. Those involved in the discussions were all familiar with the project and the normal working conditions in a dairy factory.

As always, the tendency was to imagine a great many conditions which, when considered realistically, were seen to be so unlikely ever to occur that they could be disregarded. By providing an alternative manual control for the key movements, the great majority of unscheduled conditions could be remedied, in the remote possibility of their ever occurring.

On the other hand, failure of the vacuum to hold the butter while it was being transferred to the wrapping position was a condition which, at the time, seemed to warrant a subcircuit to react automatically. Accordingly, a subcircuit was included in the main circuit which would cause the vacuum pad to lift and the traverse cylinder to return the pad and its lifting cylinder to the start position above the reception platform.

Circuit diagram in basic form

The sequence of operation had taken definite form at this stage and a circuit diagram was produced by the pneumatics engineer covering the basic form of the process. There still remained, however, a good many details which had to be determined by discussion with all involved, before the final power and control equipment could be specified. The main details to be examined included:

1. the range of speed and thrust each individual movement may be required to attain by simple adjustment or flow of pressure.
2. the physical sizes of the components resulting from the requirements determined by the decisions made in respect to speeds and thrusts.

    The question of physical size also had to be related to items of equipment which, at the time, were readily available. For example: The pneumatics engineer preferred to use double-acting cylinders

throughout because of the superior control gained. However, the folding flaps desirable from a mechanical point of view had to be of a size which could not accommodate the smallest double-acting cylinders available at that time. Single-acting cylinders of a size more compatible with the flap operation were available and were accordingly used. The basic circuit, in turn, had to be designed to accommodate these, rather than the double-acting type. Later, when suitable double-acting cylinders became available, they were incorporated in machines built since that date and the circuit was modified to cover their inclusion.

3. sensors.

These were initially selected by the pneumatics engineer on the basis of:
(a) compatibility with working environment
(b) sensitivity as related to their particular tasks
(c) repetitive consistency
(d) life expectancy
(e) simplicity of design and construction
(f) physical compatibility with overall design of machine
(g) ease of protection against possible environmental contamination
(h) flow rates and exhausting characteristics.

The initial selection was then discussed and mutual agreement reached with the mechanical engineer.

4. service units—filters, lubricators and pressure regulators.

Again these were initially selected by the pneumatics engineer and his recommendations were accepted by the mechanical engineer. In respect to the lubricators, the recommendations were initially of a tentative nature. Final decisions on lubricators were made at the trial running stages.

5. estimated cycling times.

A rough check was made and final discussions held with all involved before the pneumatics engineer was able to take the basic circuit diagram and complete the details. A circuit was then produced.

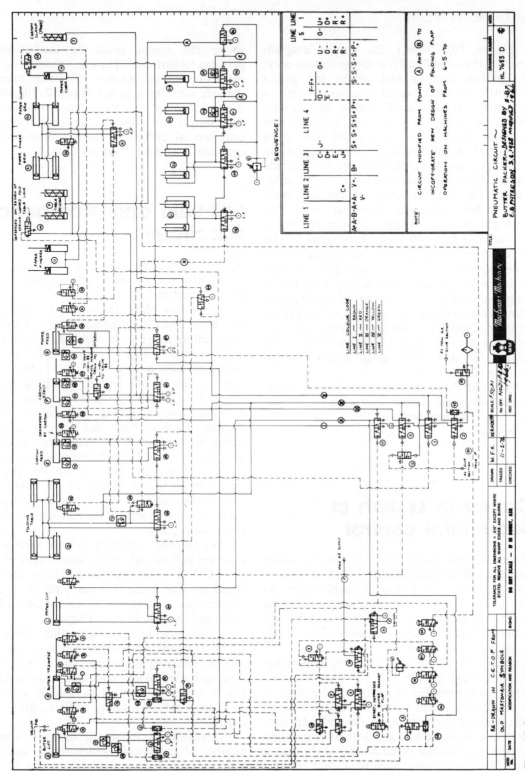

Fig. 10.2 *Butter packer pneumatic circuit diagram*

Figure 10.2 shows the circuit diagram covering the actual butter-packing section only. The carton forming and taping and gluing sections were each treated as separate sections with their circuits interlocked at appropriate stages. As will be seen from study of the diagram, the sequential operation relies on a five-group cascade system. The configuration in this diagram was that of the original cascade system. As mentioned in Chapter 2, the configuration for four or more groups has more recently been changed. However, with this particular machine, since it proved pneumatically reliable from the start, there seemed to be no good reason for changing the original circuit more than was strictly necessary. Thus the only modifications are as quoted above in respect to the use of double-acting instead of single-acting cylinders in one section. Mechanical improvements in respect to materials, bearings, etc., are naturally incorporated in the machines now being built, but the original control system remains virtually intact.

Relating the circuit diagram to some of the aspects of circuit design dealt with in previous chapters, a number of these are clearly recognisable.

Cascade system of sequential control

Obviously, the circuit relies on a base built on a five-group cascade system. Valves YY, X, Y and Z provide the five pilot supply lines. As mentioned earlier, the configuration, using four five-port valves to supply five lines, is that initially used in the cascade concept. It remained unchanged for many years. The change to the later configuration, employing one valve per supply line for four or more groups, was regarded by many as being more of an academic concept rather than one demanded from trouble experienced in the field. In fact, there is a pressure loss and loss of lubricating oil with the passage of an air signal through a valve. Logically, there must be a stage reached

where the signal could be inadequate to perform its function. Thus, in establishing rules of good common practice, it is better to err on the safe side. As seen in Chapter 2, the configuration for four or more groups—one valve for each supply line—results in air passing through only two valves from its entry as mains air to its emergence as a pilot supply line.

Fail-safe and emergency subcircuits

At first sight, there appears to be something wrong with the way in which the cascading has been carried out.

In the first place, the letters do not appear to follow completely consecutively. There are some letters missing. It should be remembered in this respect that the circuit shown here represents only a portion of the whole process. This applies to the butter wrapping portion only. Interrelated with it are the carton forming, carton gluing and carton taping sections. Each of these has its own subcircuit which in its individual design has used the cascade sequence method. However, in writing down the overall sequence of the process, letters have found themselves in other subcircuits through the close interrelationships the particular movements have had with others in other subcircuits.

The second anomaly which strikes the observer at first sight is the appearance of $A +$ $A - B - A +$ all in group I. The assumption could then be made that line I pilots all these movements and, in so doing, breaks all the rules and bounds of possibility.

Closer examination shows that this is not so. In fact, the start signal which causes A to extend, bringing the lifting pad down to the waiting butter block, emanates from a valve, No. 2, which is supplied from main air direct and forms part of a safety and emergency subcircuit.

This valve 2 is turned on only when a number of conditions coincidentally exist. To start the cycle for the first time, valve 24 must be momentarily depressed to turn on valve 2. Thereafter, once the cycle has progressed to group II movements, valve 2 is reset from direct application of line II to turn it on. This is required because, as soon as the start signal has performed its function and cylinder A has started to move down to the butter block, pressure from the rear end of cylinder A turns off valve 2. The start valve 1 cannot operate again until the machine is ready to receive a new block. The other conditions which must exist before the start signal will be possible are that cylinder B must be in its correct position above the reception platform, valve 8 must not be held in its depressed position by vacuum from the lifting pad and, of course, the flap which depressed the start valve must be held against the valve by a new block of butter.

Valves 3 and 8 ensure that these conditions are complied with. In other words, cylinder A will not extend to pick up another block of butter until the preceding block has been left on the folding table and traverse cylinder B has fully returned to its original position above the reception platform.

When cylinder A has retracted, group I supply line comes into operation, setting off the next movement, $B-$. The resetting of valve 2 then takes place after line II has become pressurised.

It should also be noted how, if the vacuum lifting pad and a butter block part company prematurely, valve 8 changes over with the loss of vacuum resulting from the vacuum pad orifices no longer being blocked by the butter. If this happens, reversal of valve 8 causes the reversal of valves 7, 9 and 10. A then will retract and B extend, bringing the vacuum lifting pad back to the start position. Here, then, is an emergency subcircuit built into the main circuit to deal with what is considered to be a feasible unscheduled occurrence in the cycle.

Valve 8, incidentally, is a diaphragm-operated valve, the diaphragm of which has a large enough area to sense the working range of vacuum required to lift the butter without sucking the butter into the vacuum orifices in the lifting pad.

Pneumatic logic

Several instances of logic conditions described in Chapter 4 will be recognised in this circuit.

One such instance is that seen at the upper centre of the diagram. Valve 60 will not allow cylinder F to operate until a particular stage has been reached in the taping of a preceding carton. Valve 83 in the taping section delivers a pressure signal to valve 60 giving the "all clear" for cylinder F to carry on. When F has completed its movement a signal is returned to the taping section to valve 88 from valve 47 in the wrapping section, allowing the taping section to carry on in its more or less independent operation.

Lower centre of the diagram, valve 15 is another example of two sets of conditions requiring to be satisfied before, in this case, the next step in the wrapping machine can proceed. The emergence of a signal from the outlet port of valve 15 indicates that a condition in the gluing section exists, a signal from valve 80 there having turned valve 15 on. The signal from valve 31 in the wrapping section, indicating that the paper has been cut, is applied to the inlet port of valve 15. Thus, with 15 open and the signal at the inlet passing through the outlet port of 15, the next step continues in the wrapping cycle, because of the evidence of compliance with the required conditions preceding that step.

Time delays: Bleed off method

The folding flap cylinders, $S1$, $S2$, $S3$ and $S4$ shown on the right side of the diagram, provide a practical illustration of a delay combined with the sensing of the exhaust decay at the end of a cylinder stroke indicating the end of the cylinder's movement. Note how, in this case, the method described in Chapter 6 of *Practical Pneumatics*, dealing with delays uses differential valves which, in themselves, are also the five-port relay valves required to operate the double-acting flap cylinders. With a constant controlled pressure on the smaller end of each differential valve, repetitive consistency in the set delays is maintained.

Changeover of valve 56 provides the signal for the folding table to open and drop the wrapped butter into the carton waiting in position below.

Before leaving this section, note that the air supply to valves 38, 40, 41 and 42 is marked "LP". Although not shown on this section of the overall process diagram, the air marked "LP" is controlled by adjustable pressure regulators. Thus, a fine degree of control can be attained over the actual folding and final pressing of the folds.

Control of movement and productive time

Cylinder movement was dealt with in detail in *Practical Pneumatics*. Examination of the cylinder control and its implications will present a more complete picture of the design features which had to be considered in the development of successful operation of the machine.

Reference has just been made to the fact that some relay valves operating cylinders have their air supply drawn from a source shown on the diagram as "LP" as distinct from the other source shown on the diagram as "HP". The LP source indicates a controlled and adjustable pressure below that of the normal mains air supply. The HP source indicates normal mains supply air taken through a pressure regulator set a few kPa below the normal minimum pressure in the fluctuating mains air supply.

Two readily recognisable examples will suffice to indicate the degree of thought which must be given to every individual movement in any project if it is to be a total success. Many a major and embarrassing catastrophe is on record which would never have occurred if the project engineer concerned had realised that every project, no matter how sophisticated it

may appear on the surface, is only comprised of a larger or smaller number of simple segments. Failure of any one of these can spell disaster for the whole project.

As two typical examples of movement control and the implications and repercussions derived from such control, cylinders A and B provide aspects of interest.

Cylinder A—butter lift
(top left of diagram)

The relay valve controlling this cylinder A is valve 4 on the diagram. Valve 4, in the first place, was selected to match its flow rate, or Cv factor, with the range of speeds at which it was considered cylinder A would ever feasibly be required to move. Pipe size, compatible with the valve size was also specified. To enable the speed in either direction to be varied up to the maximum the valve and pipe size would allow, flow regulators 17 and 18 were fitted—having flow rates equivalent to that of the relay valve 4.

Extension of A, A +, brings the vacuum lifting pad down to the butter block. On making contact, it then exerts sufficient pressure to imprint on the butter a brand design. This pressure is critical. Too much pressure can force butter up into the vacuum lifting pad, blocking the orifices and sending the butter back through to the vacuum pump. Too little pressure will not make a clear brand imprint. Thus, on the extension stroke, the cylinder has a finely controlled pressure driving the piston down, controlling the thrust, while the exhaust escapes via a flow regulator, adjusted to provide the precise speed considered desirable.

On the upstroke, or lifting stroke, the cylinder has to lift the full weight of the butter and lifting pad. This requires full lines pressure and, to control the speed, the exhaust is controlled by an adjustable flow regulator. Speed is adjusted to the maximum possible while still retaining adequate cushioning at the top of the stroke so that there is no jolting as it reaches the end of the upstroke. Excessive jolting could shake the butter block off its lifting pad.

It should also be noted that any unnecessary time taken to lift the butter on its entry into the wrapping section could reduce overall production of the plant. In this case, the lifting is a productive movement. At maximum output, the packer produces 360 blocks, wrapped and cartoned, per hour. This lifting, then, has to be repeated once every 10 seconds. An extra second of waiting for the lift to take place could reduce overall production by as much as ten per cent. However, with the low pressure drive down, the lifting movement starts almost immediately the relay valve 4 is changed over. Were the pressure to drive the piston down normal mains pressure, there would be approximately one second delay while the exhaust pressure for the upstroke escaped through the flow regulator's restriction before the exhaust pressure was reduced sufficiently to allow the cylinder to retract with its load. The cylinder will only move when the balance of forces has reached the state described by the equation $P_1 > P_2 + L$. If the exhaust pressure starts at an appreciably lower figure than that of mains supply, there is minimal delay before take-off.

This application was made because of its compatibility with the mechanical design of the flaps and the manner in which the flaps were required to perform. Lack of space precluded mechanical sensors.

The first flaps up are moved by cylinders $S1$. On completion of the outstroke of cylinders $S1$, a signal is required to outstroke cylinders $S2$, bringing into operation the second flaps and the second fold in the wrapping paper. The signal is obtained from the exhausting of pressure from the front ends of cylinders $S1$ through valve 38 at the completion of $S1$'s outstroke. Removal of pressure from the larger end of differential valve 40, causes valve 40 to change over and cylinders $S2$ to extend. At the completion of $S2$'s outstroke, loss of pressure, as with $S1$, causes valve 41 to change over and cylinder $S3$ to extend.

The fold made by cylinder $S3$ and its flap requires to be held in place under pressure for a short period before the next fold is made. Thus, rather than allow the next valve, valve 42, to change over immediately, pressure is retained for a controlled period of time on the pressure end of valve 42 and the auxiliary reservoir. The rate of flow of the air exhausting from the valve pressure end and reservoir is controlled by the flow regulator shown on the diagram as valve 43. Valve 42 changes over when pressure has dropped low enough to allow the pressure set on the other pressure end (end with a small area) of valve 42 to move the valve's spool.

As with the fold made by the flaps powered by $S3$, those powered by cylinder $S4$ are required to be held and pressurised for a controlled few moments after the fold has been made. Accordingly the changeover of valve 56 is delayed by the setting of flow regulator valve 69 after cylinder $S4$ has completed its outstroke.

Cylinder *B*

The movement transfers the butter block from the lifting position to the wrapping position. A relatively long stroke, which does nothing constructive in the further processing other than to move the block to where it can receive further attention, it might be described as an unproductive productive movement in the cycle. As such, the maximum speed is desirable as it is directly responsible for a portion of the cycling time.

Moving a block of butter suspended from a vacuum lifting pad at speed, requires both a controlled acceleration and deceleration if the two are not to part company, block and pad. The method by which this is achieved can be seen in the circuit diagram, and is as follows:

1. Valve 18 and pipes connecting it were sized for maximum feasible speed.
2. Flow regulators 28 and 29 were sized in the same manner.
3. Low, controlled pressure was provided to drive the cylinder out on its unladen, return to start, stroke.
4. Three-port valve 14 was provided, with the same Cv factor as that of valve 18, to allow the exhaust air to escape at the same rate as the driving air flowed into the cylinder, for the first portion of the stroke.

5. A sensor which could be depressed at an adjustable distance along the stroke was provided to change over valve 14 at the position best able to combine maximum speed with adequate cushioning at the end of the instroke of *B*. This sensor is seen on the diagram as valve 16.

The contribution to the end result of each of the items listed above could be described as:

Matching flow rates of relay valve 18, pipes and flow regulators 28 and 29 places no restrictions on attempts to reach the maximum feasible speed.

Low, controlled pressure on the exhausting side, after valve 18 changes over, results in minimal delay in the exhaust pressure reaching take-off point.

Valve 14, being changed over by the same signal which changes over valve 18, is open and allows the exhaust air to escape direct to atmosphere from the instant the cylinder starts to move.

Acceleration of *B* continues for quite some distance into the stroke while the piston catches up with the exhausting air and restores the balance of forces to $P_1 > P_2 + L$. By this stage, the cylinder has reached maximum speed and the stroke is well on the way to completion.

Changing over valve 14 by means of a signal from sensor valve 16 then directs the exhausting air through flow regulator 28.

By manipulating the adjustment of flow regulator 28 and the distance along the stroke at which sensor valve 16 is depressed, the correct deceleration is attained. The speed is reduced just before cylinder *B* enters its cushioning area, to approximately 0.4 metre per second—the speed at which the built-in cushioning in the cylinder can cope adequately with the mass the cylinder is moving.

Building the machine

Final drawings, mechanical and pneumatic circuits were completed. In doing so, care was given in arranging the layout from all aspects to provide ease of maintenance.

The pneumatic circuit diagram included an identification number for each individual control component. These numbers were also stamped on metal tags so that, when the control components were mounted on the machine and control panels, the numbered tags were attached to the components—each having the same number as was drawn on the circuit diagram. This later proved invaluable, not only as an easy means of describing the actual workings of the control system in the instruction booklet supplied with the machine, but also in tracing faults and in recording routine maintenance carried out on the machine from year to year. It also was found to cut down the piping-up time when the machine was being built and assembled.

The drawings were then placed in the hands of those in the workshop who would assemble the machine, or machines, which together made up the total process equipment.

Included in the workshop staff allotted to the construction of the machine were two fitters who normally specialised in the fitting and piping-up of pneumatic equipment. They had had suitable training in pneumatics, could readily interpret and work from circuit diagrams and had a full appreciation of how the equipment represented on the diagram would, or should, actually behave when connected up.

The pneumatics engineer retained close liaison with those in the workshop throughout the construction period.

Trial running of the machine in the workshop

When the machine had been fully assembled and was ready to run, it was "dry cycled", or run without butter, for five or six days. During this time, individual speeds of all cylinders were adjusted, pressures adjusted and mechanical movements checked. Likewise on the check list were such items as adjustment and positioning of sensors, positioning and accessibility of emergency buttons, fail-safe operations and their effectiveness.

One most important aspect which received attention during the trial run period was the question of adequate lubrication of all pneumatic components. Each valve was tested individually for evidence of lubrication. The test applied was a simple but effective one. Against the exhaust ports of each valve a clean piece of white paper was held while the valve performed its normal function in the cycle three or four times. The paper was then held up to the light. A slight, shiny smear showing that oil was present was the evidence sought to indicate adequate lubrication of the particular component. Thorough wetting of the paper would have shown over-lubrication, while no signs of oil at all on the paper obviously indicated that no oil was reaching that particular component.

In the case of the butter-packing section, these tests showed that the one lubricator fitted initially was not providing all components with lubrication. By the end of the trials, four lubricators of appropriate flow rates had been fitted in different positions throughout the circuit. In the carton-forming section it was found that two lubricators were required, while one was adequate for the taping and gluing section. An additional lubricator was necessary to cater for the block cutting and weighing section. Tests with the white paper held against the exhaust ports finally proved that every component in the system was adequately lubricated, by matching flow rates with demands and correct positioning of the lubricators.

The trial running period not only acted as a proving time but it also resulted in the general "bedding down" of all working components. Valve seals, initially somewhat dry, were given a chance to loosen up and acquire a fine film of lubricating oil all over so that all components worked freely and easily.

This trial run period is always worthwhile—indeed, it cannot be recommended too strongly. Any small defects or adjustments to be made can receive attention far more easily before the machine leaves the workshop floor, rather than after it has been installed in haste on a factory floor some distance away from the workshop with all its convenient plant and tools. Furthermore such attention will be far less costly than when installed on the factory floor. There, it is

an essential link in the production link and a stoppage of any kind for a minor adjustment means a stoppage on the factory production and loss of output. Too often, what are passed off as "teething troubles", which some of the less-informed treat as a necessary inconvenience associated with any new project, are simply the evidence of inattention to the preliminary essentials before the machine left the workshop. Frequently this occurs from the setting of impossible deadlines by people with no realistic knowledge of what is involved from an engineering point of view. Pressure is maintained on the workshop to cut down the time of assembly and commissioning. Such false economy usually proves exceedingly expensive. It can have a direct influence detrimental to both maintenance costs and working life of the machine.

Preparation of factory maintenance staff

During the first few years of its introduction to a market not particularly familiar with pneumatics, each butter factory about to receive a machine was requested to send an engineer to the butter packer supplier's establishment to undergo a two days' instruction course. The course covered a brief rundown on the fundamental characteristics of compressed air, the operation of air equipment and, in particular, the operation and maintenance of the packer and its ancilliary equipment.

The air supply in each factory was also inspected by the pneumatics engineer to make sure that the factory would enjoy the full potential efficiency of the machine—a matter of advantage to both user and supplier.

Installation and commissioning

The installation and commissioning of the machine was carried out by the supplier's staff under the direction of the design engineer in charge of the project. During the commissioning period, the factory staff received instruction on the operation of the machine—methods of adjusting speeds, relating of circuit diagram to control system, daily draining of filters and moisture traps, checking of oil levels in lubricators and life expectancy of components as related to routine maintenance.

All in all, attention to detail from start to finish has contributed in no small way to the enviable reputation this butter-packing equipment enjoys in the dairying industry throughout the world.

Chapter 11

Harnessing industrial vibration with compressed air

The field of vibration is complex and wide-ranging. The effects of vibration on matter subjected to it are equally so. Although an area not fully explored, it is gaining increasing recognition as a major factor for good or ill in the industrial scene. As its nature becomes more recognisable, and its effects in different contexts more predictable, it will more easily be made to work for, rather than against, the industry it serves. Effective vibration control has a very direct bearing on the degree of success achieved in higher standards of goods and services, and a higher quality of life for both those engaged in production and those who are the recipients of the goods and services resulting from productive effort.

Although the picture is incomplete, it is known that vibration can be both a friend and an enemy. The following are typical, authentic cases of uninhibited transmission of vibration being an enemy.

Evidence that uninhibited vibration destroys buildings

1. An oil-fired, electric power generating station of some size has as part of its complex a multi-storeyed building housing heavy machinery. At ground level, six large piston type compressors, mounted on very large concrete blocks, transmit to their surroundings disturbing frequencies of 570 cycles per minute (cpm) or 9.5 cycles per second (hertz or Hz). The concrete blocks do little to prevent the transmission of this low frequency. In a period of two and a half years, this vibration has attacked the main supporting columns of the building so successfully that cracks up to 10 mm wide are visible in a number of places. Without the remedial action incorporating the application of compressed air in the form of Airmount® isolators, the machinery housed

131

on all the upper floors would have spectacularly come down to ground level. To quote one of those involved in the problem, "More get at-able there, but not quite so tidy".

2. A corrugated cardboard factory uses a large hydraulic shredder to recycle waste board and trimmings from the process. It is mounted on concrete of considerable depth. To gain better quality control, a closed circuit TV is mounted on the first floor, checking continuously on a section of the processing. From this, measurements are transmitted to a central control. Although the camera was situated some distance from the shredder on the floor below, the measurements were inaccurate and the camera broke down regularly. The cause of this was vibration transmitted through the building from the shredder. When investigating the cause of the troubles experienced by the camera, it was discovered that the whole first floor concrete floor was badly cracked all over the whole area. Not only did the first floor house the camera, but also many tonnes of paper—records, documents of various types and sundry invoice books for branches, etc. The building was found to be in a dangerous condition, primarily caused by vibration from the shredder.

Effect of vibration on instruments

1. A laboratory in a meat research institute employs an electronic microscope to measure tiny sections of animal tissue. Vibration from a highway and a railway, both one to two kilometres distant from the laboratory, destroyed any possible accuracy of measurement. Thus, until the instrument had been isolated from the low frequency vibrations travelling through the ground from the various sources, any research work carried out which relied on information from the instrument in question was totally valueless. In this case the traditional methods—rubber blocks, springs and concrete mass had all proved ineffective.

2. A dairy factory made extensive alterations and additions to its plant during what is called the "shut down" period—the two months or so in which cows in New Zealand are not producing milk. At such a time, any major changes in plant are made. In this case, so much had to be done in the limited period available that there was no time in which to test the plant properly before the milk started to arrive from the farms. Once the supply starts, it continues to do so twenty-four hours per day, seven days per week for the next ten months. When the whole plant started up, the engineers were appalled to find that the stainless steel benches in the laboratory were visibly shivering in response to vibration transmitted through the building from various types of compressors, pumps and fans processing the raw materials. All offending machines had been mounted in the traditional ways known at the time to the plant designers. However, with the low frequencies put out by these machines, the methods used were of little or no value. Mounting machinery on Airmount isolators was a method outside the experience of those involved.

Plans were prepared for a building in which to house the worst of the offending machines, which would then be located on the far side of the site, some distance from the factory and its laboratory. Pipes were planned to run from machines to factory underground. The cost of such remedial action was quite enormous. However, before any such work was actually started, by chance the chief engineer heard of the application of compressed air in isolating low frequency vibration. Mounting one of the worst offenders, a compressor working at 570 rpm, on three Airmounts, achieved isolation of 98.2 per cent. To do so took only a matter of hours. With such proof of what could actually be achieved, they lost no time in applying this method throughout.

The dairy factory laboratory referred to not only provides important day-to-day information on the total operation of the plant, but the information acquired is applied in their research program seeking both better quality of existing products and also ways of diversifying into other end products from the common raw material. Only when all instruments could work without interference from vibration could such information be of any value.

Thus there is ample evidence to indicate that unless vibration is thoroughly controlled, delicate instruments will not produce reliable results and, in turn, research work and quality control can be useless.

Mechanical maintenance

1. A paper mill had occasion to shut down for four months one of the several paper machines employed in the plant. The other machines continued to run for the normal twenty-four hour, seven days per week operation.

   At the end of the period, it was found that, before the idle machine could be started up again, all the bearings had to be renewed. They had been completely worn out, not by any work that the machine had done, but by vibration transmitted through the floor from those machines still working. It does not take much imagination to realise that much of the normal mechanical maintenance in any plant can be eliminated by effective isolation of vibration transmitted from one machine to another.

2. A sand plant pumping sand from a river bed employs a vibrating screen weighing some ten tonnes to screen all the materials flowing from the pumps. It is located at the top of a high building in which the rest of the plant separates the gravel, sand and pumice as it flows by gravity down through the building. The screen for many years was mounted on steel springs. Eventually the noise created throughout the building by this particular screen was found to have serious permanent effects on the hearing and health of those employed there. The noise was so great that anyone shouting beside the screen was inaudible to anyone else standing alongside. Standard Airmounts, each provided with additional volume, isolated the screen

by over 98 per cent. The noise was reduced to the level in which it is now possible to stand alongside the screen and converse in normal tones. Only after this isolation was achieved did the plant engineer realise that most of the time had previously been spent in picking up the nuts and bolts falling off the rest of the machinery and screwing them back in place. This "normal" maintenance work had kept several fitters busy all day and every day. Plant improvements and modifications had always, till then, had to take second priority just to keep the plant running.

Thus, there is ample evidence that mechanical maintenance in a plant in which vibration is not properly controlled is very much greater than need be. Vibration transmitted from one machine to another will wear out the second machine even though the second machine may do no work at all.

Uninhibited vibration

Medical research in recent years into the effects of noise and vibration on the human frame has revealed disturbing facts in respect to personnel associated with manufacturing industries. The results of such research are well documented and published. From the evidence produced by such researchers, it is clear that vibration not only affects the senses—sense of touch, optical and audio—but also the response to such sensors. Thus, the imperfect messages from affected sensors are conveyed in a distorted fashion to the brain. The time taken to convey such messages is also longer than normal.

People affected by vibration may suffer physical damage, sometimes temporary, sometimes permanent; sometimes moderate, sometimes fatal. They become accident prone, irritable, impatient. Interpretation of such things as figures displayed on instruments to assist them in controlling a process can be distorted and unreliable.

In the broader context, the effects of vibration on industrial personnel are far reaching, with a direct bearing on the economy of the industry in respect to loss of production, quality control, absenteeism, strained industrial relations and damage to both plant and personnel.

A practical example may be quoted. A new machine was installed in a factory and the operator of the machine instructed in its use and control. Within a few days, the operator complained at intervals through the working day of feeling sick. He would often disappear into the toilets for periods of up to twenty minutes or so, leaving his machine either shut off or unattended. As the weeks went on, it was noted that every third or fourth day he was absent from work.

Prior to the installation of the machine, he had been one of the most reliable employees in the place. On the days when the regular operator was absent, those taking his place also complained of feeling sick after an hour or two on the machine.

Finally, the supervisor of that factory section decided to work the machine himself for a day or two to test the truth of the claims made by those who had previously done so. Within two or three hours, he was overcome with nausea. After a brief spell away from the machine, he recovered. Returning to the machine, the nausea came back in a short time.

The cause of such effects on the operators was finally attributed to the vibration emanating from the machine. The machine was then effectively isolated and the problem disappeared.

Thus, there is ample evidence to prove that uninhibited vibration can cause serious inefficiency in respect to personnel as well as machines in industry, with equally serious effects on the overall economy of the industry in question.

Successful application of vibration

On the other hand, vibration is applied successfully in many industrial processes. Vibration is commonly used in:

1. the screening of materials—sand, coal, road metal, peas and beans to name a few;

2. the transfer of materials such as milk powder, small components in the assembly of manufactured products, sand and so on;
3. stress relief of metals prior to machining to avoid the distortion of metals which sometimes can take place during the machining of such metals.

In addition to all this, the practical engineer has usually found from bitter experience that the simple rule-of-thumb methods of isolating vibration which he was taught in his early apprenticeship do not work very well when applied to the lower range of frequencies so common in modern industry.

There was a time when it was considered sufficient to mount troublesome machinery on a concrete block which in itself weighed three times the weight of the offending machine. There are instances of this which can be quoted where vibration from machines isolated in this way on cubes of concrete buried seven or eight metres into the ground can be picked up ten kilometres away.

Steel springs are often used as vibration isolators. For frequencies above 1000 cycles per minute, or 16.6 Hz (when isolation of 95 per cent or more is required), they can prove very effective. However, for frequencies below 1000 cpm, they prove increasingly difficult to apply as the lower the disturbing frequency, the longer must the spring be to attain the required effective natural frequency in the mass/spring system for any worthwhile isolation result. Problems of stability intrude until the steel spring becomes impractical for the purpose.

Rubber of varying degrees of hardness and flexibility is often used effectively in the higher frequency ranges. But for disturbing frequencies in the lower ranges, as with steel springs, rubber has severe limitations. On the other hand, a flexible airbag provides an isolating spring of an entirely different nature.

The concept of isolation with compressed air was expressed as early as the 1860s. Patents have been taken out covering ideas which might be described as the embryo stages of the present concept. Very little more appears to have been done to develop the idea until the 1930s when the Firestone organisation investigated the potential of such a concept as a means of protecting loads carried on heavy transport road and rail cars. The success of this company's activities in transport encouraged their research engineers to expand the concept to embrace vibration generally throughout the industry.

The action of an airspring, or, as they are commonly termed, Airmount isolator, when dealing with vibration from a machine mounted on it, may be described as follows. The Airmount, as a rule, has a top and bottom metal plate between which and attached to which is the flexible bellows. Figures 11.1 and 11.2 show two commonly used types.

Vibration from the machine attacks the top plate in the form of a continuing series of impacts. These impacts are transformed on the underside of the top plate into a continuing series of air shock waves in the compressed air inside the Airmount. The volume of air in the Airmount acts as a surge tank, dissipating a proportion of the shock waves before they reach the bottom plate. If there were sufficient volume there, these shock waves would be almost wholly absorbed or dissipated before they reached the lower end of the Airmount. Only a very small proportion would remain, transmitted mainly down the flexible walls of the Airmount. The materials used in the flexible walls are such as to ensure this leakage is so small as to be scarcely worth taking into consideration. If the volume in the Airmount is such as to allow relatively high proportion of the shock waves to reach the lower end plate, still more of the force remaining in them can be removed by adding more volume. This is done by making up a reservoir, or small pressure tank, and connecting it to the Airmount, either directly through an orifice or by means of a flexible tube. The residual shock waves will then pass on to the auxiliary surge tank to be further dissipated. Normally the final amount of force in the disturbing frequency or vibration passes out through the base plate again as a series of impacts. What passes through will be a proportion of the initial force on the top plate. This residual force is described as a percentage of the original and is said to have

been transmitted through the Airmount, e.g. the Airmount under the specific working conditions is said to have a transmissibility of, say, 3.4 per cent. Or, looking at the reverse side of the picture, the machine is said to have been isolated in respect to the disturbing frequency it creates by 96.6 per cent, or isolation of 96.6 per cent has been attained.

If any airspring is compared with a conventional metal spring, it will be found that whereas with most conventional metal springs the spring rate is a constant, the airspring rate is not. The rate of the Airmount is a function of the change of effective area, volume and pressure from the design height when vibrating under a given load. Hence Airmounts are often described as variable rate springs.

As stated at the outset, vibration and vibration analysis is a complex field. For those who wish to acquire an in-depth knowledge, there are a number of good publications on the subject. Several such books are listed at the end of this chapter. A subject closely allied to vibration is noise—an equally complex area. In both, to carry them to the ultimate, it is necessary to enter the realms of higher mathematics in a manner seldom associated with the day-to-day work of the busy industrial engineer, who is occupied with the daily running of an increasingly sophisticated industrial process and the need to maintain ever increasing production output. Because of all these factors, average practical engineers tend to avoid vibration problems, if possible. Because their predecessors ignored such problems, even classified them as the inevitable accompaniment of some types of equipment, they will tend to adopt the same attitude and "leave well alone".

On the other hand, the cost of allowing uninhibited vibration to run rife through a plant in terms of loss of efficiency of plant and personnel, high maintenance, poor quality control, inaccurate records and research data and reduced life expectancy of all plant and buildings is so high that no industry can afford to ignore vibration problems wherever they may occur.

The purpose of this chapter is not to attempt another textbook in depth on the subject, but rather to provide the busy engineer

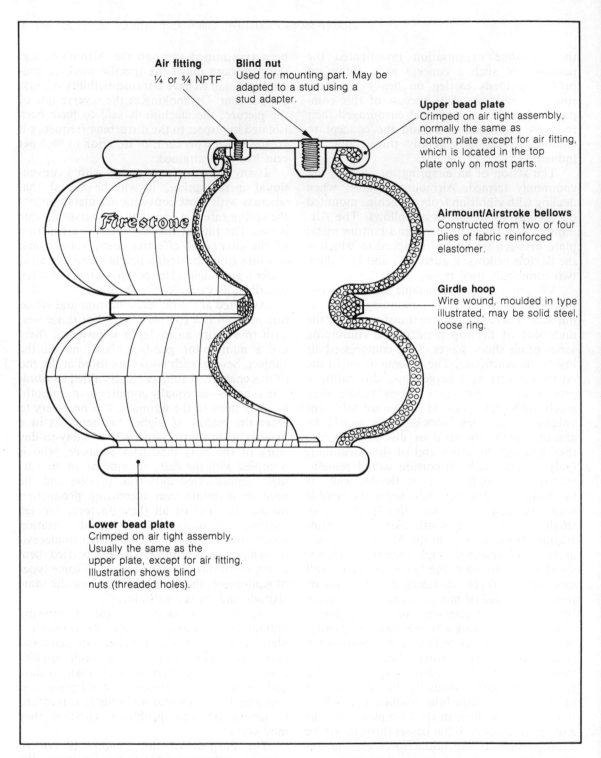

Air fitting
¼ or ¾ NPTF

Blind nut
Used for mounting part. May be
adapted to a stud using a
stud adapter.

Upper bead plate
Crimped on air tight assembly,
normally the same as
bottom plate except for air fitting,
which is located in the top
plate only on most parts.

Airmount/Airstroke bellows
Constructed from two or four
plies of fabric reinforced
elastomer.

Girdle hoop
Wire wound, moulded in type
illustrated, may be solid steel,
loose ring.

Lower bead plate
Crimped on air tight assembly.
Usually the same as the
upper plate, except for air fitting.
Illustration shows blind
nuts (threaded holes).

Fig. 11.1 *Convoluted style air spring*

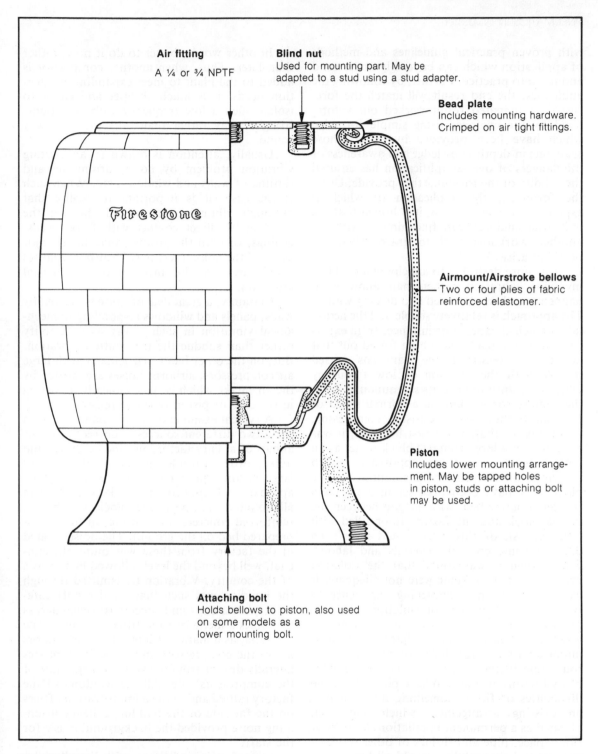

Air fitting
A ¼ or ¾ NPTF

Blind nut
Used for mounting part. May be
adapted to a stud using a stud adapter.

Bead plate
Includes mounting hardware.
Crimped on air tight fittings.

Firestone

Airmount/Airstroke bellows
Two or four plies of fabric
reinforced elastomer.

Piston
Includes lower mounting arrangement. May be tapped holes
in piston, studs or attaching bolt
may be used.

Attaching bolt
Holds bellows to piston, also used
on some models as a
lower mounting bolt.

Fig. 11.2 *Reversible sleeve type air spring*

with proven practical guidelines and methods of application which can be readily understood and put into practice. Further, by following the guidelines, the end result will match the forecasted, anticipated result worked out beforehand by the engineer using simple formulae which have been provided by those whose academic in-depth knowledge and awareness of the dangers of over-simplification has ensured the validity of the formulae they provide. Using the formulae, the application of which is explained on pages 142–6, it is hoped that the industrial engineer can find himself with yet another workable tool to grasp when the occasion arises.

It is more desirable to anticipate a problem involving vibration rather than allow it to happen and then be forced into dealing with it. The approach is relatively simple and the action taken much cheaper. For instance, from experience it may already have been found out that an air compressor in the plant has caused trouble with the creation of low frequency vibration transmitted to its surroundings from the compressor working at a normal speed of 570 rpm. In the case of the existing compressor, it is probable that, when first installed, it was mounted on a large concrete block to act as an "inertia block" which was supposed to absorb the vibration created by the compressor. The block may have even been set in a pit on a further concrete bed, with the gap between the inertia block and the factory floor filled with bitumen. All of this work would have been done at some cost of materials and labour. Then, when it was found that the isolation properties of the scheme were not adequate, it would have been a chastening experience to find that the concept of isolation by compressed air not only existed, but was simpler, less costly than the original method and, most important of all, provided an isolation factor of something better than 95 per cent. Experience of that nature would also have pointed out the difficulties of fitting something different into an existing arrangement which had been planned as a permanent installation. Matters of floor space, pipe and fittings of other services, and general accessibility, would all have come to light and presented problems of their own.

In other words, better to do it now, rather than later. Thus, when another compressor is added to the plant to meet expanding production needs, it is much cheaper and easier to isolate such a low frequency with Airmounts straight away rather than take the long way round.

Usually, attention is drawn to an existing vibration problem by noise—drumming and rattling of walls and windows, etc. When such is the case, it is important to realise that although vibration transmitted through the building by direct contact with floor, walls, ceilings, or even the ground, may be a major cause of the noise, there may be airborne noises which are not the product of mechanical vibration.

Certainly, a great deal of noise is created by walls, panels and windows responding to mechanical vibration in such a way as to amplify rather than subdue the transmitted vibration. But, on the other hand, particularly with large air compressors, airborne noises are created by the air intake which is spasmodically occurring in the case of piston type compressors.

A typical example of noise emanating from these two different sources is on record in the annals of a dairy factory producing butter and milk powder. Included in the plant are three large piston type compressors, each weighing approximately four tonnes. Initially, these were all mounted on heavy rubber blocks on a heavy, reinforced concrete floor sitting, in turn, on a prepared bed on the ground. The noise created in the factory from these was quite phenomenal, well beyond the levels allowed by the laws of the country. Vibration transmitted through the building was such that, in the staff cafeteria, plates, cups and saucers travelled across the tables a metre or so in fifteen minutes. The cafeteria, as a matter of interest, is three floors above the compressors and some forty metres laterally distant from the vertical height line of the compressors' site. All the windows of the factory rattled and in the administration offices on the far side of the building a heavy drumming noise provided the background music for the staff.

In such a dairy factory, with literally kilometres of stainless steel pipe connecting the maze of different types of stainless steel vats, pressure vessels and hoppers, such vibration cannot be allowed to run rife through the plant. Apart from the discomfort of the staff, stainless steel is brittle and vulnerable to vibration. Food processors find the repairs and maintenance to stainless steel equipment a large item on their maintenance budget. In this case, the mounting of the compressors on Airmounts successfully isolated the mechanical vibration. The drumming noises in walls, ceilings, floors, etc., disappeared. No longer did the crockery walk around the tables in the cafeteria. The noise level fell dramatically, but not completely. The windows still rattled. The cause of this was finally traced back to the compressor intakes. There, large regular volumes of air were being sucked in at high speed as the pistons rose and fell. This set up air pulsations which travelled around the outside of the building in a remarkable fashion, causing the windows to rattle in response to the pulsating pressure changes outside. This problem was then dealt with by building a form of maze through which the intake air passed to the compressors.

The point, of course, is that while isolation of mechanical vibration plays a major part in noise level reduction, it should not be considered as the only possible contributory factor.

Aspects to be reviewed in dealing with a vibration problem

The types of vibration to be isolated may be either created for a purpose, such as screening, material transfer or metal stress relief, or it may be inherent in a machine performing some other function.

Inherent in the machine

This can be treated as a straightforward isolation problem following the guidelines described later in this chapter. It should, however, be remembered that often vibration in a machine may not be inherent but rather due to faulty lubrication, poor selection of bearings, or worn parts. Therefore, before attempting to assess the amount of vibration emanating from such a machine, these points should be checked and any faulty parts replaced or repaired. Excessive vibration is often the warning given by the machine before a complete breakdown.

Thus, the normal degree of inherent vibration should be assessed. Then often it is desirable to attach instruments to the machine which will sound the alarm if the vibration rises above the normal level. Otherwise, the excessive vibration may go unnoticed.

Induced vibration

Vibration may have been created for a specific purpose. In such a case, care should be taken to ensure that, in isolating the vibration, the absorption of the forces by the Airmounts is not detrimental to the process for which the vibration has been created.

For example, a vibrating screen may be designed in such a way that the heavy base on which it is mounted may be required to reflect or bounce back the forces created by the vibrator. Thus, rather than mount the screen on Airmounts, the screen and its base will need to be mounted on the Airmounts. Cases of concrete block-making machines have been known in which the performance of the machines suffered when attempts were made to isolate the machines and so prevent vibration from the vibrators being transmitted to the rest of the plant. In such cases, the vibrators were relying on a solid base to act as an anvil bouncing the forces back into the product as it formed in the moulds. The obvious remedy was to mount the machines on heavy reinforced concrete inertia blocks, which in turn were mounted on Airmount isolators. The same detrimental effects on the efficiency of a vibratory feeder could occur if the overall design is not studied first to ascertain how the designer planned to transmit the forces created by the vibrator into the product or material being transferred.

139

The same thought should apply to vibration applied in the stress relief of metals. Prevention of transmission to the surroundings of the source should not interfere with the process for which the vibration has been created.

The degree of isolation required

The degree of isolation required should be assessed before any attempt is made to select a suitable type of Airmount. This assessment will place the isolation factor in one or other of three categories:

1. the highest possible isolation against a range of frequencies;
2. an isolation factor of above 95 per cent for frequencies of 300 cpm and up;
3. isolation of machines creating impact disturbing frequencies.

Those problems in the first category are usually those associated with delicate instruments in a laboratory. To maintain maximum accuracy in performance, the maximum isolation has to be achieved against all possible sources of vibration—people walking by on the footpath outside (120 cpm), heavy traffic on the nearby highway (300 cpm), machinery in the building itself such as air-conditioning fans (600–3000 cpm), compressors (570–1450 cpm), pumps (300–1000 cpm), and so on. As will be seen when discussing the working formulae used in selecting suitable Airmounts, the lowest disturbing frequency is taken as the target. If adequate isolation is achieved from the lowest disturbing frequency, isolation from the higher frequencies will, as a general rule, be greater. Sometimes harmonics (maybe only the first) need to be considered.

Where the higher possible degree of isolation is of primary importance, the type of Airmount used will more likely be the reversible sleeve type as in Figure 11.2. Natural frequencies of 20 cpm or lower can be obtained by arranging suitable combinations of mass, type of Airmount and additional volume. The working formulae to be described later in this chapter apply to this type of work in the same fashion as they do to the general run of the mill isolation cases. However, the characteristics of

this type require careful consideration in relating them to the mounting details, permissible ratio of lateral to vertical stability, working conditions when in use, and so on. To be certain of the best possible end result, it is well to consult someone experienced in the field.

In selecting the reversible sleeve type rather than the convoluted type, special attaching metals are required to achieve lateral stability, i.e. the rubber part is enclosed in a cylindrical metal sleeve. Thus in designing the system, experience in matching the normal working conditions with the characteristics of the system can be helpful. Suppliers specialising in Airmounts usually provide a technical back-up for such occasions.

However, where such advice is not immediately available, the following guidelines will be found useful.

1. The metal sleeve encasing the outside of the rolling sleeve rubber part is necessary to maintain lateral stability (see Fig. 11.3).
2. The sleeve length and the ID can be varied to influence the amount of lateral stiffness needed for the application.

Fig. 11.3

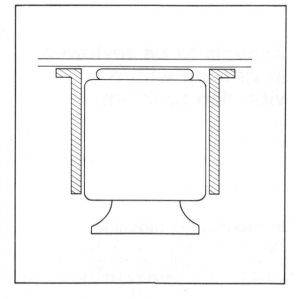

3. Vertical natural frequencies for such systems can be brought down below 60 cpm using additional volume by connecting a reservoir to each.
4. Height control air valves will be required for most such applications.

Isolation of machinery employed in production

This will include all manner of moving devices which tend to send out vibration or disturbing frequencies, covering a wide range of speeds (300 rpm–3000 rpm).

For such equipment, the single or double convoluted type will be the more suitable type of Airmount as a general rule. When inflated to correct static design height and any pressures above 275 kPa (40 psi), they provide remarkably rigid mounting devices with satisfactory lateral stability for a wide range of working conditions.

Machines creating impact disturbing frequencies

Typical machines of this type would include drop-hammers, large power presses, guillotines, etc., which deliver intermittent heavy blows or impacts rather than a continuous series of regular vibrations.

Isolation of the forces transmitted from such equipment is usually more effectively achieved with Airmounts than by any other means. On record is the successful isolation by Airmounts of a steam forging hammer which, together with its base inertia block, weighed over one million kilograms or one thousand tonnes.

So successful was the isolation in this case that a glass of water filled to the brim and placed on the ground less than a metre from the block upon which the machine was mounted showed no sign of any kind that the forge hammer was pounding away alongside.

Successful isolation of this type of slow moving equipment calls for careful consideration of a great many different factors including the machine manufacturer's basis of design.

For example, one well-known brand of power press is made in three sections, mounted one above the other, and held rigidly in line by long tie bolts firmly attached to a heavy, rigid base. The base serves the twofold purpose of reflecting back the primary press action, anvil fashion, and of holding the tie bolts in such a way as to hold the machine in precise alignment.

The general method of selection and mounting on Airmounts is to treat the problem as one of cushioning a force or impact, rather than the absorption of continuous vibration. The forces involved are calculated on the basis of newton metres or inch tons and then selection of the Airmount made on the basis of acceptable movement. Restraint of the movement is often provided by Airmounts applying pressure to friction pads on the sides of the inertia block or by attaching restraints in the form of more conventional shock absorbers. The greater the use of such restraints, the greater the transmission or leakage of the disturbing force to its surrounding environment.

It is important to remember that, if the power press is working at a regular speed of, say, 60 rpm, the natural frequency of the Airmount incorporated in the spring-mass system selected should not be at the same 60 cpm. Where the disturbing frequency and the natural frequency of the spring-mass system are of the same frequency, the one excites the other, producing considerable motion.

Where it is not possible to attain a natural frequency of a mounting system significantly below that of the disturbing frequency, the natural frequency of the system must be raised to a figure above that of the disturbing frequency. There are times when inadvertently coincidental frequencies may occur and it is then necessary to raise the Airmount natural frequency. This may be done by adding water to the Airmount in sufficient volume to reduce the volume of air and, in turn, raise the natural frequency to a more desirable level. To estimate the volume of air which must remain and the natural frequency attained thereby, the formula described later which establishes the spring rate, K, will serve to do so.

Usually, when the correct type of Airmount has been selected, they are placed under the inertia block around the outside edges or four corners. Mechanical restraints above, below and on all four sides, placed, say 25 mm from the block itself when the Airmounts are inflated to design height, provide safety against earthquake or malfunctioning of air valves. The general rule of "minimum distance between mounts to be no less than twice the vertical height of the centre of gravity from the top of the Airmount when inflated" holds in these cases as in all others. This ensures the necessary lateral stability of the system.

As with the cases mentioned above, it is worthwhile having the design of the proposed system checked by an engineer experienced in this form of application to ensure that all factors which might affect the end result have been taken into consideration.

Airmounts for this purpose can be applied to any size of plant with certainty of success provided all relevant information is obtained and considered at design stage.

While all three general classifications of vibration and shock isolation are common enough in any industrial scene, by far the greater of the three in productive industry is that classified on page 141 as "Isolation of machinery employed in production". In this group occur most of the problems which beset modern industry. Vibrations put out by machines working at speeds of anything from 300–3000 rpm include pumps, compressors, fans, generators, vibratory feeders and shaker screens, cyclones, saw and a host of special purpose machinery peculiar to their own industry. Until recently, much of the vibration emanating from such machinery was accepted as inevitable. However, the awareness of the damage caused by such vibration is not only growing among those engaged in industry itself, but also among those who legislate and the government departments concerned with the overall industrial efficiency and the welfare of those working in industry.

Thus, most countries are laying down more stringent regulations in respect to what they consider acceptable levels of noise and vibration.

To anyone unacquainted with vibration isolation using compressed air, the levels specified appear to be almost unattainable when relating to frequencies of 1000 cpm or lower. However, as the ensuing pages will show, the correct application of Airmounts provides answers well within the limits of the most stringent regulations.

Before attempting to select a suitable Airmount, the following relevant details will be required from the machine to be mounted:

    overall weight
    weight distribution over base
    vertical height of centre of gravity
    overall general dimensions
    normal working speed or speeds
    location of machine
    source and description of disturbing
        motions and forces
    function of machine
    type of work performed.

The following formulae constitute the basis of correct selection and forecasting of the result which will be attained.

Where the spring rate is known, the following formula will establish the natural frequency of the isolator under a given load condition:

$$fn = 188 \sqrt{\frac{K}{L}}$$

where K = dynamic spring rate (lb/in)
       L = load (weight) (lb)
       fn = natural frequency, cpm (cycles per minute)

e.g. where the spring rate is 415 lb/in and the load 1080 lb

then $fn = 188 \sqrt{\dfrac{415}{1080}}$

        = 116.538 cpm or 1.94 cycles per second (hertz)

With a known natural frequency this can be matched against the disturbing frequency to establish the amount of vibration which escapes, or is transmitted, through the isolator. In turn, the percentage of isolation can then be established. To do this, the following formula provides an answer:

142

$$I\% = 100 - T$$

$$T\% = \frac{100}{\left(\dfrac{F_f}{F_n}\right)^2 - 1}$$

where

I = % of isolation
T = % of transmission
F_f = force or disturbing frequency
F_n = natural frequency of isolator

Example

F_f Force frequency   780 rpm
F_n Natural frequency 116.538 cpm

$$\therefore \quad T = \frac{100}{\left(\dfrac{780}{116.538}\right)^2 - 1}$$

$$= \frac{100}{44.798 - 1}$$

$$= 2.28\%$$

$$\therefore \quad I = 100 - 2.28$$

$$= 97.72\%$$

The dynamic spring rate itself is established by a formula developed by Charles William Grepp in the course of research carried out by the Firestone Industrial Products Company's engineering research and development group. His formula was developed to combine in a single expression several steps which begin with the polytropic gas law. The Grepp formula gives a very close to real-life result for all conditions of area and volume and for excursions of approximately plus and minus 10 mm.

$$K = (P_g + 14.7)\left[A_c\left(\frac{V_1}{V_c}\right)^{1.38} - A_e\left(\frac{V_1}{V_e}\right)^{1.38}\right]$$

$$- 14.7\,(A_c - A_e)$$

where K = spring rate (lb/in)
P_g = gauge pressure at design height (psi)
A_c = effective area at -0.5 inches from design height (in²)
A_e = effective area at $+0.5$ inches from design height (in²)
V_1 = internal volume at design height (in³)
V_c = internal volume at -0.5 inches from design height (in³)
V_e = internal volume at $+0.5$ inches from design height (in³)

To establish the value of each of the above, the data covering the characteristics of the isolator must be consulted. Figure 11.4 shows a typical data sheet provided by the Firestone manufacturers. Note that typical examples are given to save time in making up a short list of types when making a selection.

Figure 11.5 shows how to arrive at the individual values for the formula using the data sheet. Effective area is the reactive area needed to support a given load at a specific pressure. Unlike pneumatic cylinders, Airmount isolators and Airstroke® actuators are non-linear devices and their effective areas vary with changes in height. To determine effective area (A) at a given height use the following formula:

$$A = \frac{L}{P_g}$$

where A = effective area
L = load
P_g = gauge pressure

To illustrate consider a practical example:

A Number 22 Airmount isolator has been selected to support a 4000 lb load. We can read the following information directly from the data curve:

design height (HT) = 9.5 inches
volume   (V)   = 780 cubic inches
load   (L)   = 4000 lb
pressure   (P_g) = unknown
effective area (A) = unknown

In the example the effective area (A) and the required pressure (P_g) must be calculated. From the chart it is apparent that the pressure will be between 60 and 80 psi. Effective area is relatively constant at any given height and will vary only a minor amount under different pressures. By using the pressure curve closest to the actual operating conditions the effective area can be calculated as follows:

$$A = \frac{L}{P_g}$$

$$= \frac{4250}{80}$$

$$= 53.125 \text{ in}^2$$

Since our example has a load of 4000 lb instead of 4250 lb one additional calculation is needed to determine the pressure. Use the revised formula:

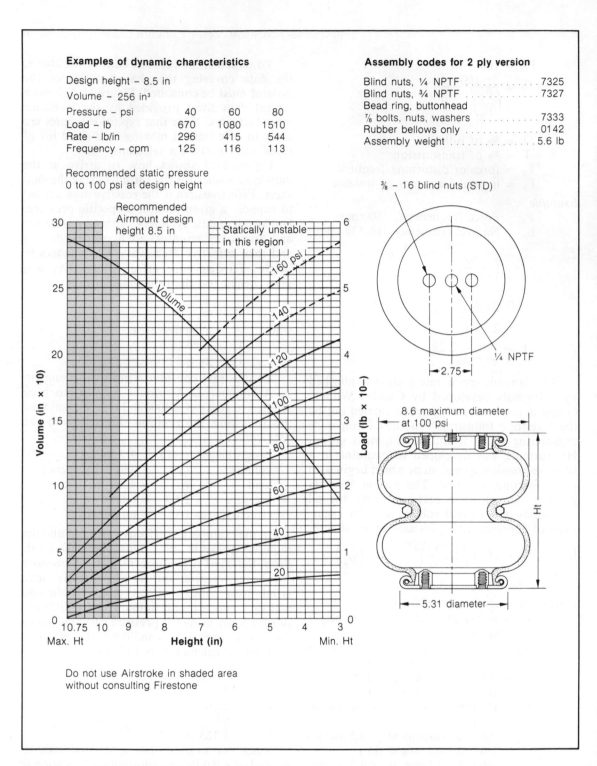

Examples of dynamic characteristics

Design height – 8.5 in

Volume – 256 in³

Pressure – psi	40	60	80
Load – lb	670	1080	1510
Rate – lb/in	296	415	544
Frequency – cpm	125	116	113

Recommended static pressure
0 to 100 psi at design height

Recommended
Airmount design
height 8.5 in

Statically unstable
in this region

160 psi
140
120
100
80
60
40
20

Volume

Volume (in × 10)

Load (lb × 10⁻)

30 25 20 15 10 5 0

6 5 4 3 2 1 0

10.75 10 9 8 7 6 5 4 3
Max. Ht   Height (in)   Min. Ht

Do not use Airstroke in shaded area
without consulting Firestone

Assembly codes for 2 ply version

Blind nuts, ¼ NPTF	7325
Blind nuts, ¾ NPTF	7327
Bead ring, buttonhead ⅞ bolts, nuts, washers	7333
Rubber bellows only	0142
Assembly weight	5.6 lb

⅜ – 16 blind nuts (STD)

¼ NPTF

2.75

8.6 maximum diameter
at 100 psi

Ht

5.31 diameter

Fig. 11.4 *Typical characteristics data sheet*

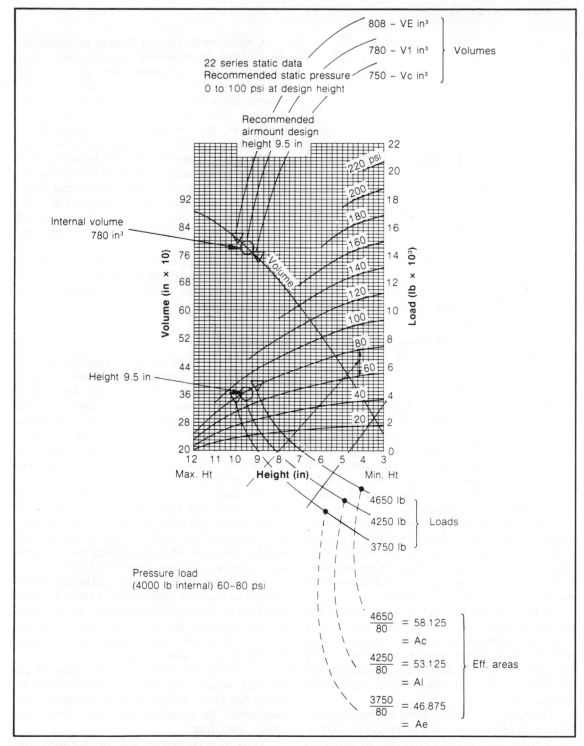

Fig. 11.5 *Method of establishing formula component values*

$$P_g = \frac{L}{A}$$

$$= \frac{4000}{53.12}$$

$$= 75.3 \text{ psi}$$

$$K = (P_g + 14.7)\left[A_c\left(\frac{V_1}{V_c}\right)^{1.38} - A_e\left(\frac{V_1}{V_e}\right)^{1.38}\right]$$

$$- 14.7 (A_c - A_e)$$

$$= (75.3 + 14.7)\left[58.125\left(\frac{780}{750}\right)^{1.38}\right.$$

$$\left. - 46.875\left(\frac{780}{808}\right)^{1.38}\right]$$

$$- 14.7 (58.125 - 46.875)$$

$$= 90\,[58.125(1.0556) - 46.875(0.9525)]$$
$$- 14.7(11.25)$$

$$= 90\,[61.3568 - 44.6484] - 165.375$$
$$= 90\,[16.708\,4] - 165.375$$
$$= 1503.756 - 165.375$$
$$= 1338.381 \text{ lb/in}$$

The typical examples of dynamic characteristics, however, do not show the change of spring rate which can be brought about by adding volume. While the natural frequency at design height will show very little change with change of load condition and related pressure, change of volume can show quite dramatic changes in natural frequency.

While one type may appear to have a lower natural frequency than another with no additional volume, it may not be so when volume is added to each. Thus, when selecting, it is worthwhile examining several alternatives before making the final decision.

When considering additional volume there are two possible approaches. One is to decide upon a desirable minimum isolation figure and, with all other factors known, work backwards through the above formula to establish the unknown figure for the volume required to achieve this.

The other way, which from a practical point of view is often more convenient, is to look at the feasible in terms of availability of materials and cost of making up additional reservoirs.

A paper by B. Dorion-Brown, published by the CSIRO of Australia, suggests that when adding volume, an appreciable gain can be attained up to a ratio of 3:1—added volume: volume already in isolator at design height. Beyond this ratio the gain progressively diminishes. The implication is that beyond this ratio it is not worth pursuing additional gain.

However, when the problem is approached from a practical angle, it is often found that for little cost a reservoir made from scrap short lengths of steam pipe may give a ratio of 1:10. The additional volume over the 1:3 ratio will be found well worthwhile in reducing the transmission.

Two practical examples will show how these formulae fall into place in the selection of the most suitable type of Airmount isolator for a particular application.

Example 1

A 200 f³/min (5.66 m³/min) capacity, piston type compressor complete with motor mounted on a common base is to be installed. The overall weight to be mounted is approximately 1305 kg or 2871 lb. The compressor will have a working speed of 645 rpm and the motor 1440 rpm.

The centre of gravity of the combined compressor, motor and base will be relatively high, approximately one metre.

The base is of heavy RSJ (rolled steel joist) construction and rectangular shape, 1662 mm by 1105 mm. The concrete floor in the compressor room has been laid to support this sort of weight with ease when distributed over mounting plates with a total area of 0.27 m².

The above details provide sufficient information to enable a selection of a suitable type. Looking at the weight distribution on the base, it seems obvious that approximately two-thirds of the total weight is at the compressor end and one-third at the motor end.

In the interests of keeping a near equal load on each mount and the ever present need to spend no more money than is necessary, the load could be supported by three Airmounts rather than four.

Thus the catalogue will be scanned for an Airmount capable of supporting a load of

957 lb with a pressure preferably between 40 psi and 80 psi. Also the type to seek will have the lowest possible natural frequency to cope with the relatively low disturbing frequency of 645 cpm, emanating from the compressor at 645 rpm. If the isolator can isolate the lowest disturbing frequency, it will provide even greater isolation for the higher disturbing frequency emanating from the motor at 1440 rpm.

Scanning the manufacturer's catalogue, the data sheet covering a no. 26 Airmount shown in Figure 11.4 appears to provide a suitable answer.

The typical dynamic characteristics show that for a load of 1080 lb at a pressure of 60 psi the natural frequency is 116 cpm.

Matching this against the disturbing frequency the formula below is used:

$$T = \frac{100}{\left(\frac{F_f}{F_n}\right)^2 - 1}$$

In this case:

$$T = \frac{100}{\left(\frac{645}{116}\right)^2 - 1}$$

$$= \frac{100}{30.9 - 1}$$

$$= 3.34\%$$

This may be considered adequate for the purpose but it is worth realising that this transmission can be considerably reduced with added volume to each isolator. While the installation is a "one off" matter and the cost of reservoirs when compared with the cost of the compressor is minimal, whatever vibration is allowed to escape will continue to do so over the years to the detriment of other plant and building.

Thus the feasible type of reservoir is investigated. Short lengths of large bore steam pipe appear to be available at little cost. A length of 8″ ID bore × 34.5″ (200 mm ID bore × 895 mm) would provide, when sealed at each end, approximately one cubic foot which could be added to the volume already in the Airmount.

To establish the spring rate with the added volume of 1728 cubic inches, the appropriate formula is put to work.

$$K = (P_g + 14.7)\left[A_c\left(\frac{V_1}{V_c}\right)^{1.38} - A_e\left(\frac{V_1}{V_e}\right)^{1.38}\right]$$
$$- 14.7\,(A_c - A_e)$$

Or, when adding volume to each Airmount by connecting to each a reservoir, the formula should be written:

$$K = (P_g + 14.7)\left[A_c\left(\frac{V_1 + V_R}{V_c + V_R}\right)^{1.38}\right.$$

$$\left. - A_e\left(\frac{V_1 + V_R}{V_e + V_R}\right)^{1.38}\right] - 14.7\,(A_c - A_e)$$

where V_R = reservoir volume in cubic inches.

Referring to Figures 11.4 and 11.5, the values for the above can be established.

To establish the value of the working pressure, one or two points are worth noting. Since the effective area at design height varies slightly on the pressure changes, the following method will give a remarkably realistic answer. By taking the load supporting capacity on the pressure curve above that of the particular instance and also the load carrying capacity on the particular pressure curve below the particular instance in question, then averaging the two, an answer would be obtained with a minimal error. In this case, the weight to be supported on each Airmount is 960 lb. Therefore:

1. At 8½″ design height at 60 psi the load supported is 1080 lb at 8½″.
   Therefore, the effective area equals 18 in² at 8½″.
   At 40 psi the load supported is 670 lb at 8½″.
   Therefore, the effective area equals 16.75 in² at 8½″.
   Average area between these two pressures
   $$= \frac{18 + 16.75}{2}$$
   $$= 17.375$$
   Therefore, the pressure to support 960 lb
   $$= \frac{960}{17.375}$$
   $$= 55.25 \text{ psi at } 8\frac{1}{2}″$$

147

2. The values for A_c and A_e will also need to be determined by much the same method. From the data sheet of the no. 26 Airmount, we can obtain the following details:

At height of 8″, 60 psi supports 1200 lb.
At height of 9″, 60 psi supports 950 lb.
Therefore,

A_c at 60 psi = 20 in² at 8″.
A_e at 60 psi = 15.83 in² at 9″.

At height of 8″, 40 psi supports 750 lb.
At height of 9″, 40 psi supports 570 lb.
Therefore,

A_c at 40 psi = 18.75 in² at 8″.
A_e at 40 psi = 14.25 in² at 9″.

Therefore,

average A_c between 60 and 40 psi

$$= \frac{20 + 18.75}{2}$$
$$= 19.375 \text{ at } 8″$$

average A_e between 60 and 40 psi

$$= \frac{15.83 + 14.25}{2}$$
$$= 15.04 \text{ at } 9″$$

Now our values to establish K spring rate are as follows:

P_g = 55.25
A_c = 19.375
A_e = 15.04
V_1 can be read from the data sheet as 250 at 8½″
V_c can be read from the data sheet as 240 at 8″
V_e can be read from the data sheet as 260 at 9″

Since additional volume is to be added to each Airmount, the additional volume must be added to each of the volume values. Thus,

V_1 = 250 + 1728 = 1978 at 8½″
V_c = 240 + 1728 = 1968 at 8″
V_e = 260 + 1728 = 1988 at 9″

The spring rate K can now be determined by inserting the values for this particular case.

$$K = (55.25 + 14.7) \left[19.375 \left(\frac{250 + 1728}{240 + 1728} \right)^{1.38} \right.$$
$$- 15.04 \left. \left(\frac{250 + 1728}{260 + 1728} \right)^{1.38} \right]$$
$$- 14.7 \, (19.375 - 15.04)$$
$$= 69.95 \left[19.375 \left(\frac{1978}{1968} \right)^{1.38} \right.$$
$$- 15.04 \left. \left(\frac{1978}{1988} \right)^{1.38} \right] - 14.7 \, (4.335)$$
$$= 69.95 \, [19.375 \, (1.007 \, 019)$$
$$- 15.04 \, (0.993 \, 065)] - 63.724 \, 50$$
$$= 69.95 \, [19.510 \, 993 - 14.935 \, 697]$$
$$- 63.724 \, 50$$
$$= 256.317 \, 45 \text{ lb/in} \quad \text{i.e. } 256.32 \text{ lb/in}$$

With a known spring rate of 256.32, the natural frequency can be established as follows:

$$f_n = 188 \sqrt{\frac{K}{L}}$$
$$= 188 \sqrt{\frac{256.32}{960}}$$
$$= 188 \, (0.516 \, 720 \, 4)$$
$$= 97.143 \, 435 \quad \text{i.e. } 97.14 \text{ cpm}$$

The transmission now through the Airmount with its extra volume in its reservoir will be:

$$T = \frac{100}{\left(\dfrac{F_f}{F_n} \right)^2 - 1}$$
$$= \frac{100}{\left(\dfrac{645}{97.14} \right)^2 - 1}$$
$$= \frac{100}{44.088 \, 287 - 1}$$
$$= 2.32\%$$

Isolation is now in the order of 97.7 per cent, a figure which should prove satisfactory.

Accordingly, the decision is made to mount the compressor on three no. 26 Airmount/Isolators, working at a static design height of 8½″ and a pressure in the order of 55 psi.

Mounting

Simplicity and stability are the two main considerations when mounting the machine.

Simplicity

Desirably the least possible number of Airmounts should be used. This will depend on the strength of the frame supporting the machine and the weight-carrying capacity of the floor on which the machine is situated.

Fig. 11.6 *Three mounts to match weight distribution*

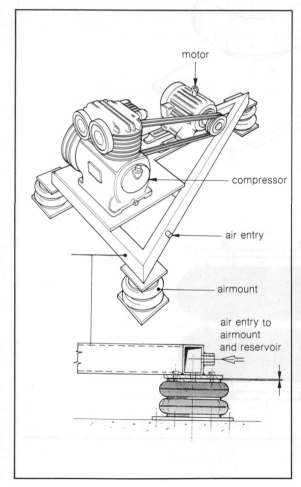

motor

compressor

air entry

airmount

air entry to airmount and reservoir

Sometimes fans, compressors and pumps will have a weight distribution over the base, in which two-thirds of the overall weight is on one end of the base and one-third on the other. In such cases, three Airmounts will usually prove adequate. This then gives a more equable load per Airmount and less equipment, less cost and less maintenance (see Fig. 11.6).

Stability

This will depend on both the distance between mounts as related to the vertical height of the centre of gravity and the working pressure of the Airmounts themselves.

The normal rule of thumb method of determining the minimum distance between mounts has already been described as twice the vertical height of the centre of gravity. However, it is obvious that the nature of the forces—vertical, lateral, rotary—put out by the machine itself when working will determine to what extent this method will suffice, e.g. machinery in which dynamic forces are all vertical or near vertical will allow a ratio of almost 1:1 rather than 1:2, centre of gravity vertical height/distance between mounts. Other types of machinery may have lateral forces of such magnitude that side-stabilising Airmounts will be required to restrain movement within acceptable limits. In all cases, pressures of 40 psi to 80 psi are desirable to attain a reasonable rigidity and stiffness of the air spring. It should be remembered that no matter what the normal working speed may be, when the machine in question is started up it will run through all the range of frequencies from zero to its normal cycling speed. Thus, for a brief period, the natural frequency of the Airmount and the machine's rising frequency will coincide. At that point, movement will occur until the machine's frequency rises above the critical coincidental frequency of Airmount and machine. Airmounts with a stiffness stemming from pressures above 40 psi will generally limit the momentary movement to no more than a few millimetres.

Where the centre of gravity is low and central enough to be contained within the boundaries of a line drawn around an Airmount, i.e. with no overhang, brackets may be

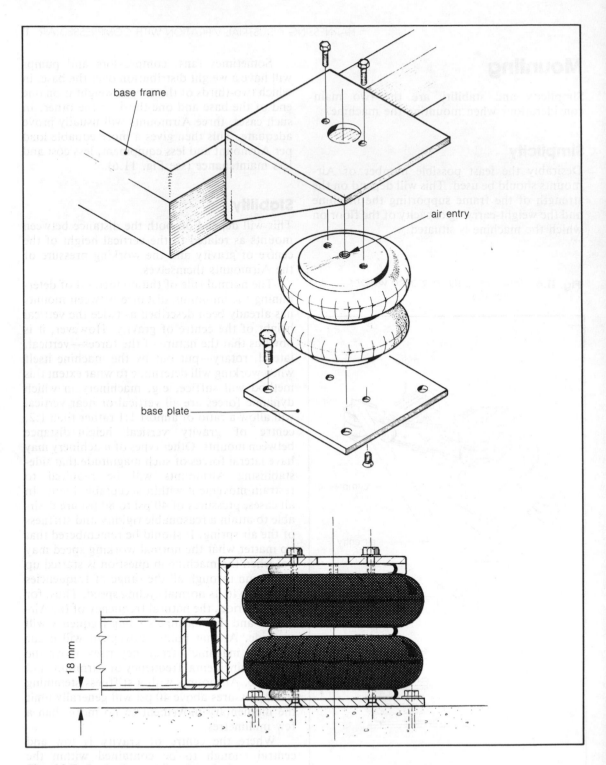

Fig. 11.7 *Typical mounting brackets*

attached to the machine base. Figure 11.7 shows typical such brackets. This would be constructed and attached to the base in such a manner as to protrude outward horizontally to provide a means of attaching it to the top plate of the Airmount. The height of such a bracket would be such as to allow the Airmount, when inflated to its recommended static design height, to lift the base of the machine by 18 mm or ¾″ above the ground or floor on which the machine normally stands.

The top plate of the Airmount must be bolted to the support bracket. The bottom plate of the Airmount is normally bolted to a steel plate which, in turn, is bolted to the floor. Provision for connections to the air entry port of the Airmount must not be overlooked.

Where the centre of gravity is high there are two options in mounting:

1. Where floor space is available, a subframe may be constructed on which the machine will be mounted. The subframe must be large enough to provide adequate distance between mounts.
2. The Airmount may be raised to meet the centre of gravity by attaching Airmounts to the top of pedestals placed around the machine. Figure 11.8 illustrates this concept. The pedestals may be constructed of hollow steel, such as steam pipe or hollow square section. If constructed of such materials, the pedestals can then serve the dual purpose of acting as a pedestal for the Airmount and also acting as a reservoir to provide additional volume for the Airmounts it supports. The height of these pedestals will be such as to bring the ratio of vertical height of centre of gravity to distance between mounts to the limits described above. The pedestals must be bolted to the floor and the Airmounts bolted both to the pedestal and the machine which they support. Air access between pedestal and Airmount is usually provided by constructing a flanged top for the pedestal. The top can then be arranged so that a matching entry hole through the top can be gasketed and fitted in an airtight manner to the Airmount plate and its air entry.

Valving

Where a constant air supply is available, the Airmounts are best served by providing permanent connections to the mains air supply. Since each Airmount will have a slightly different load as a general rule, it will require an individual pressure to maintain the standard height required all round. Thus, each Airmount should be provided with a small ⅛″ BSP pressure regulator. Since the height, rather than the pressure, is the final governing factor, no gauges are required on these small regulators. All regulators should then be connected by small bore pipe to a common manifold, which should then be connected through a non-return valve, filter and shut-off isolating valve to the main air supply. Figure 11.9 shows a typical valving assembly.

Where no air supply is available, each Airmount will require a "tank" valve. The tank valve is similar to those used in automobile tyres. The Airmounts must then be inflated by a foot pump and checked at regular intervals so that any drop in height can be rectified by air replenishment.

Where load conditions vary and Airmounts must maintain a constant design height, valving will require to accommodate the change of pressure needed to maintain the constant height. Such conditions are typically associated with vibrating, filling and weighing. The vibration is introduced to compact the material flowing in a container. The weighing is performed by the usual electronic weighing device and cannot be accurate unless the weighing equipment is protected from the vibration which compacts the material. Thus, Airmounts will isolate the vibrating container from the scales.

Initially, the Airmounts will be simply supporting an empty container on a relatively low pressure. Then, as the container is filled and the weight increases, the pressure in the supporting Airmounts will naturally have to be increased.

Figure 11.10 shows how this can be achieved. The figure shows four supporting Airmounts being used to support the load. It

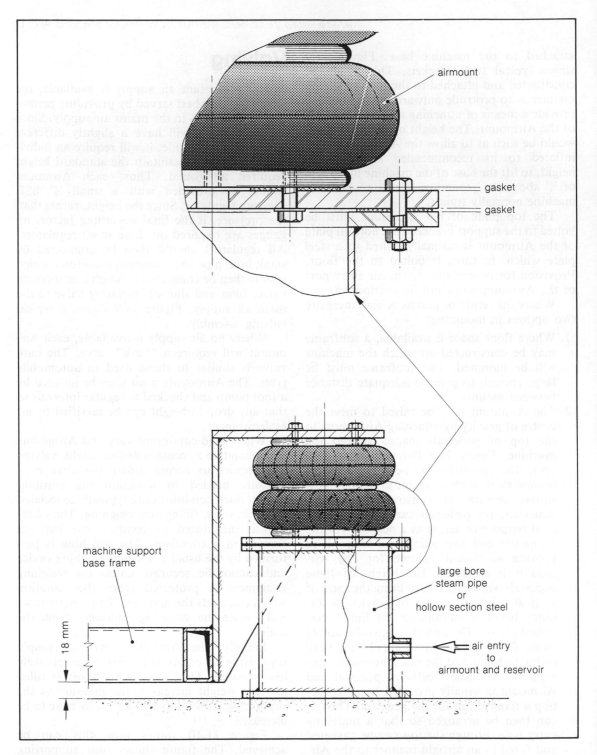

Fig. 11.8 *Mounts raised to meet high centre of gravity*

airmount

gasket

gasket

machine support
base frame

18 mm

large bore
steam pipe
or
hollow section steel

air entry
to
airmount and reservoir

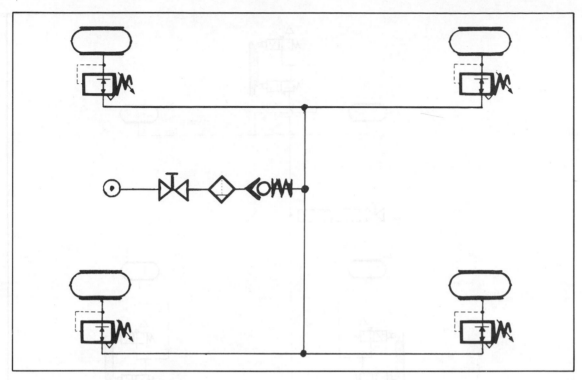

Fig. 11.9 *Typical valving diagram*

will be noted that two Airmounts are connected to the one air supply and will be working to a common pressure. The other two will each have individual height-sensing devices and individual pressure supplies. This arrangement has been proved as a satisfactory method of avoiding "hunting". The sensing devices are small pneumatic sensors, often roller operated spring return microvalves of an on–off nature. These can be either pocket valves or three-port valves with one port blocked. There is a wide variety of these offering in the pneumatic field. Because the volume of air required to either inflate or exhaust is usually fairly small, ⅛″ BSP ported valves are normally preferable, as too large a valve can tend to provide an injection of too large a volume before the injection is cut off when the design height is reached. The speed of filling, and also the time allowed for exhausting when the filled and weighed container is removed, must be taken into account

when selecting these sensing valves, since the rates of flow of these valves combined with the volume in the Airmount will be directly related to the time taken to effect any desired change of pressure. Careful study of these factors will be amply repaid in the overall performance of the Airmount system under working conditions.

There is no intention of presenting a complete picture of the complexities of vibration and noise here. This chapter is solely an attempt to provide the practical engineer with a useful tool to use when the occasion demands.

By following the guidelines described here, most isolation problems can be dealt with easily and satisfactorily. It should be mentioned that when faced with a vibration and noise problem, experience has shown that often it is better to isolate the vibration at the source before spending money and time on sound-absorbing panels and materials. By isolating the source of

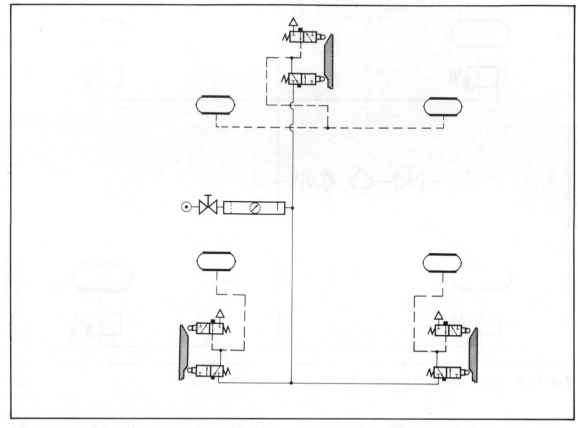

Fig. 11.10 *Valving diagram to maintain constant height under variable load conditions*

vibration first, a considerable reduction in noise level usually occurs. Vibration transmitted through a building draws responses from materials, panels, walls, beams, etc., which vary in different areas of the building. In turn, noise is the product in such cases of materials responding to the disturbing frequencies transmitted from the source. Much of this type of noise will disappear when the source has been correctly isolated.

The author is indebted and grateful to the Firestone Industrial Products Company and, in particular, Mr C.W. Grepp, for much of the material used in this chapter, together with permission to use extracts, data sheets, charts and drawings from their publications.

For those who wish to pursue the subject of noise and vibration in depth there are many publications available. Some of the more helpful are listed below.

Shock and Vibration Handbook, edited by Cyril M. Harris and Charles E. Creede, Published by McGraw-Hill Book Company, New York, 1976.

Elements of Vibration Analysis, Leonard Meirovitch, Published by McGraw-Hill Book Company, New York, 1975.

Theory of Vibration with Applications, William T. Thomson, Published by Prentice-Hall Inc., New Jersey, 1981.

Mechanical Vibrations, by Francis S. Tse, Ivan E. Morse and Roland T. Hinkle, Published by Allyn and Bacon Inc., Boston, 1963.

Vibration Analysis, by Robert K. Vierck, Published by Thomas Y. Cromwell Company Harper and Row, New York, 1979.

Check list when mounting a machine on Airmounts

Stability relies on an initial appreciation of the vertical and lateral forces produced by the machine to be mounted when it is working. This appreciation is then lined up alongside the following general rule-of-thumb assumptions.

1. The ratio of stability, lateral to vertical, at pressures above 30 psi will be from 1:3 upwards. At pressures above 55 psi the ratio will be from 1:2.
2. The ratio of vertical height of centre of gravity to minimum distance between Airmounts should never be less than 1:1 and preferably not less than 1:2. This is normally achieved either by ensuring the base is wide enough or by mounting the Airmounts on pedestals of some sort.
3. A minimum of overhang outside the lines drawn between Airmounts should be maintained.

If it is not possible to stay within the limits of these three factors, consideration must be given to the addition of Airmounts as side restraints.

Having established the requirements in terms of number and positions of Airmounts needed to create satisfactory isolation, consideration must be given to the selection of satisfactory flexible connections to those lines connected to all extraneous parts of the system, e.g. water lines, air intake and discharge lines, electrical lines, etc. These should be flexible enough to accept such movement as may take place on the part of the machine to be isolated. While such movement will be very small while the machine is running (say 1 or 2 mm), the movement can be greater when the machine starts up and runs through the range of frequencies from zero up to its normal design speed. In some cases, a momentary movement of up to 10 or 12 mm can occur as the disturbing frequency put out by the machine coincides with the natural frequency of the Airmount while its speed is building up to normal working speed.

Assembly and fitting

1. Ensure that when all flexible connections have been made no pressure is being exerted in any direction, lateral or vertical, on the machine when it is at design height (i.e. when the Airmounts are inflated to the design height recommended), e.g. flexible lines such as water lines and high temperature air discharge lines should be horizontal rather than vertical.
2. Brackets on the machine base to which the Airmounts are to be attached should be constructed in a manner which will allow adequate clearance all round the Airmount when it is inflated. A clearance of 25 mm is desirable all round.
3. The machine should be mounted on mechanical stops, blocks or the floor in such a manner as to allow it to rise a further 15–20 mm only when the Airmounts are inflated to working pressure.
4. The Airmounts should be connected to base plates which can, in turn, be bolted to the floor or structure on which the machine is to be mounted.
5. When the top plates of the Airmounts have been attached to the machine, the bottom plates must be bolted to the floor in such a manner as to ensure that top and bottom Airmount plates are precisely one above the other.
6. Air connections are better if they are made to the bottom rather than the top of the Airmount since they will not be subject to vibration which can make them work loose over a period of time. Often this is not easily arranged, in which case a little Locktite on the fitting will help to hold it leak-proof for a reasonable period of time.
7. When connecting to a main air line, an isolating on/off valve is desirable followed by a non-return valve before the manifolds from which individual lines can be taken to the individual pressure regulators controlling the height of each Airmount.
8. When the Airmounts have been bolted into position, top and bottom, the Airmounts may then be inflated to raise the machine 15 to 20 mm up to working static design

155

height. Each Airmount will probably require a slightly different pressure from that of its companions since each will be supporting a slightly different load. The important thing is that each Airmount should be brought up to its static design height and that when this is done the machine should be perfectly level. If the floor is uneven, it may be necessary to build up the height of one or two Airmounts with spacer plates.

It is important that the air pressure be introduced to each Airmount gradually in steps. When tightening up a number of nuts holding one work piece together the torque on each nut is increased in gradual stages, nut by nut, rotating round the nuts until they have all been brought up to their required torque. In the same way the pressure on each Airmount should be increased by a small increment rotating from Airmount to Airmount. After each

Fig. 11.11 *A compressor installed on airmounts: mass of machine 5 tonnes, disturbing frequency 736 cpm (12 Hz), isolation 94.7%*

airbag

bottom plate of airmount

mounting pedestal/air reservoir

increase in pressure a short period should be allowed for the effects of the change of setting of the regulator to show in the Airmount before moving on to the next regulator.

In this way the machine will be brought up in level fashion until design height has been attained at all mounting points. The inflation exercise can take up to fifteen minutes. Failure to follow out this procedure usually results in the Airmount requiring least pressure to rise first throwing the whole machine out of line.

When all Airmounts are at final design height the pressure regulators can be locked in their individual settings and the system should require no further attention other than the normal random check at intervals to see that all is well.

Protection of airmounts

Environmental working conditions can be such as to warrant giving a little thought to the protection of Airmounts from excessive dust, grit, sawdust, corrosive liquids and the like. For instance, sawdust can accumulate and work in between the flexible bellows and the metal base plate causing unnecessary wear on the flexible bellows through continuing abrasion. Often a simple cowelling of canvas, plastic or metal can prevent such materials from falling around the Airmounts.

Appendix

1. Conversion tables

The following table establishes the relativity of the various units of measurement of pressure which existed prior to the introduction of the pascal unit.

Pressure

Units	kPa
1 psi	6.894 76
1 in Hg	3.386 39
1 in water	0.249 08
14.696 psi	101.325
1 atm	101.325
1 kgf/cm²	101.325

Atmospheric pressure

14.696 psi = 101 325 Pa
           = 101.325 kPa
           = 1.013 25 bar

psi to kPa

psi	kPa	psi	kPa
1	6.894 76	40	275.790 4
2	13.789 52	50	344.738
3	20.684 28	60	413.685 6
4	27.579 04	70	482.633 2
5	34.473 8	80	551.580 8
6	41.368 56	90	620.528 4
7	48.263 32	100	689.476 0
8	55.158 08	110	758.423 6
9	62.052 84	120	827.371 2
10	68.947 6	130	896.318 8
20	137.895 2	140	965.266 4
30	206.842 8	150	1034.214

2. Units of measurement

Volume of a gas

The volume of a gas, measured in Imperial units in terms of cubic feet, is now universally measured in units of either cubic metres (m^3) or cubic decimetres (dm^3). Frequently the more popular term for decimetres, "litre", is used.

The unit, whether m^3 or litre, refers to air at 101.325 kPa (abs) at 15°C, and is the common denominator to which all matters of volume are reduced, e.g. the air consumption of an air cylinder, no matter at what particular pressure it may be working, will be expressed in terms of either litres or m^3 of atmosphere at 15°C.

```
1 cubic metre (m³)
        = 35.314 7 cubic feet (ft³)
1 litre (L)  = 0.035 314 7 cubic feet (ft³)
1 cubic foot (ft³)
        = 0.028 317 cubic metres (m³)
        = 28.317 cubic decimetres (dm³)
        = 28.317 litres (L)
```

Flow of a gas

Those accustomed to using the Imperial system of measurement until recently described the volume of air flowing through a pipe or orifice in terms of cubic feet of atmosphere per minute (ft^3/min or cfm). As with volume, the atmosphere referred to was the standard reference, air at 101.325 kPa (abs) at 15°C.

Now the universal term and unit is that of either a cubic metre or a litre of atmosphere at 101.325 kPa (abs) at 15°C per second. These are written as m^3/s and L/s.

```
1 ft³/min  = 0.000 471 947 m³/s
           = 0.471 947 dm³/s or L/s
1 L/s      = 2.1189 ft³/min
           and 1 m³/s = 2119 ft³/min
```

Force and pressure

The following explanation of these units starts with the unit of pressure and works backwards, largely because the pascal is usually the first unit about which one is asked.

1. The term *pressure* is merely a kind of shorthand for the expression *force per unit area*, whatever the units of force and area that may be used.
2. In the Imperial system (and in technical metric as well) there was no special name for the unit of pressure. Thus its name, whether pounds per square inch, kilograms-force per square centimetre, or whatever, was self-explanatory.
3. In SI the unit of pressure, the *pascal*, is only a special name for the SI unit of force-per-area, viz. the *newton per square metre*.
4. The *newton* is the SI unit of force, and is defined as: "that force which, when applied to a mass of one kilogram, produces in it an acceleration of one metre per second per second in the direction of the force."

   Because it is defined in this way, without any reference to gravitational force either on the earth or anywhere else, the newton is described as an *absolute* or *non-gravitational* unit of force.
5. *Size of the newton.* It has been found by experiment that on the earth's surface the acceleration due to gravity is approximately 9.81 m/s^2. Thus by simple proportion it follows that the force of gravity on a mass of 1 kg must be 9.81 newtons. Hence the old metric unit, the kilogram-force, must be equal to 9.81 newtons.
6. Further, since there are 2.205 pounds in one kilogram, the *pound-force* equals:

$$\frac{9.81}{2.205} = 4.45 \text{ newtons}$$

From this it can be seen that the SI unit of pressure, being 1 newton per square metre, is very small. As a result, ordinary pressures of a few psi are in SI a few hundred kilopascals. For example, atmospheric pressure of about 14.5 psi is approximately 100 kPa, while a tyre pressure of 24 psi is about 165 kPa above atmospheric.

7. The fundamental equation connecting force, mass and acceleration still holds regardless of the units used, and we can still say that:

$$F = m.a$$

where, in SI,
F is the force in newtons
m is the mass in kilograms, and
a is the acceleration in m/s^2

Note: It must be remembered that the expression "pounds per square inch" should really be given as "pounds-force per square inch" but this is usually contracted by omitting the word "force".

3. Desirable maximum piston rod length in metres for various combinations of thrust and rod diameter

Piston rod diameter (mm)	Thrust on rod (kN)															
	0.22	0.44	0.67	1.1	1.7	3.1	4.4	6.2	8.0	10.6	14.2	17.8	22.2	26.7	35.6	44.5
16	1.70	1.50	1.34	1.09	0.94	0.76	0.68	0.61	0.58	0.48	0.40	0.33	0.22			
25.4		2.79	2.61	2.38	2.10	1.72	1.52	1.34	1.21	1.14	1.04	0.96	0.87	0.76	0.66	0.53
35			3.70	3.40	2.99	2.66	2.33	2.08	1.90	1.70	1.60	1.52	1.42	1.27	1.14
44.5				4.72	4.26	3.93	3.60	3.22	2.89	2.61	2.38	2.21	2.08	1.93	1.77
50.8					5.13	4.82	4.42	4.06	3.68	3.80	3.02	2.79	2.59	2.36	2.26
63.5					6.98	6.52	6.19	5.84	5.41	4.92	4.44	4.14	3.86	3.48	3.17
76						8.38	7.82	7.51	7.13	6.62	6.09	5.71	5.28	4.77	4.36
89							9.77	9.29	8.81	8.35	7.87	7.34	6.96	6.22	5.63
101.5								11.17	10.54	10.16	9.60	9.14	8.68	7.87	7.08
114									12.39	11.70	11.32	10.82	10.41	9.52	8.86
127										12.54	12.09	11.35	10.46

Example

For a cylinder having a stroke of 1.2 m and a thrust of 8 kN, locate the 1.2 m length in the 8 kN column of thrusts, reading left to the first column, piston rod diameters, and the rod diameter is found to be 25.4 mm.

161

4. Selection of valve size for cylinder speed

Cylinder bore size (piston diameter) (mm)	Air consumption per mm movement (litres)	Valve port size	Rate of flow (litres per second)
32	0.0059	⅛" BSP	1.8
50	0.0146	¼" BSP	11.5
63	0.0227	½" BSP	37.75
80	0.0335	¾" BSP	85
100	0.0602	1" BSP	103

All above based on supply air at 700 kPa; pressure drop across valves 40 kPa

Example

An air cylinder with a bore diameter of 80 mm and stroke of 300 mm is required to move at an average speed of 300 mm per second.

From the above table, 1 mm of movement of an 80 mm cylinder requires 0.033 48 litres; therefore for 300 mm movement the air required is 0.033 48 × 300 litres = 10.0441.

From the above table, ¼" BSP valve flow is 11.5 litres per second.

Therefore, the valve selected will be the ¼" BSP.

For a cylinder with half the bore diameter of a listed cylinder, divide the consumption figure of the listed cylinder by 4. Therefore,

consumption of a cylinder 200 mm diameter bore will be:

0.0692 × 4 = 0.2768 litres per mm stroke;

consumption of a cylinder 25 mm diameter bore will be:

$\frac{0.0146}{4}$ = 0.0037 litres per mm stroke.

5. Typical cylinder thrusts

Cylinder bore diameter (mm)	Thrust (kN) working pressure 700 kPa	working pressure 550 kPa
25	0.3489	0.2741
50	1.3956	1.0965
63	2.1795	1.7124
80	3.1358	2.4638
100	5.5822	4.3859
125	8.718	6.8497
160	12.5656	9.8728
200	22.3329	17.5469
250	34.939	27.4516
300	50.3069	39.5261

To arrive at *theoretical pull* of cylinders for normal practical purposes take above figures and deduct 10%.

6. Sizing of a cylinder applied to a lever

When applying a cylinder to a lever the cylinder will need to be mounted on what is termed a trunnion mount—i.e. the cylinder body will be mounted on a hinged mounting, while the end of the piston rod will be attached to the lever by means of a "clevis", a hinged device which is shown, with the rest of the arrangement commonly adopted, in Figure A.1.

As may be seen in Figure A.1, the end of the piston rod, in following the lever to which it is attached, will traverse through an arc. In doing so, the effective force at the point of applica-tion, i.e. the end of the lever, will vary through-out the whole movement. Because of this, it must be taken into account when sizing the cylinder selected for the work. The following table, combined with the details shown in Figure A.1 provides a means of finding the effective force at the point of application.

As will be seen in the table, the force is progressively reduced as the angle of thrust from the cylinder progressively diminishes from 90° to the lever arm—angle A in Figure A.1.

Fig. A.1

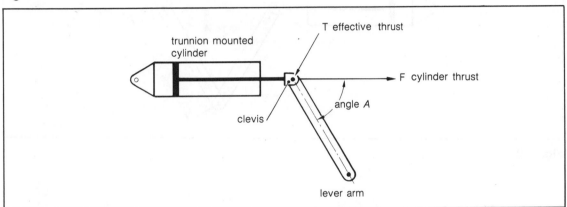

Power factors

Angle A (degrees)	Power factor (sin A)	Angle A (degrees)	Power factor (sin A)
5	0.087	50	0.766
10	0.174	55	0.819
15	0.259	60	0.867
20	0.342	65	0.906
25	0.423	70	0.940
30	0.500	75	0.966
35	0.573	80	0.985
40	0.643	85	0.996
45	0.707	90	1.000

Example

A cylinder with a bore diameter of 50 mm has a theoretical thrust of 1.084 kN at 550 kPa. It is to be applied to a lever in such a way that the most acute angle A between the cylinder and lever axis will be 50°. What will be the minimum effective thrust of the cylinder?

From the table, at 50° the effective force T will be:

$$1.084 \times 0.766$$
$$= 0.8303 \text{ kN}$$

163

7. Determining length of cylinder stroke applied to a lever

If the cylinder is rotating the lever through an arc so that the angle A is divided equally by the perpendicular line drawn from the lever's pivot point, the length of stroke may be determined by using what is termed the "chord" factor. If A is divided unequally, the arrangement will have to be drawn to scale and physical measurements will provide the answer. In the case of equal angles the length of stroke may be easily determined by taking the length of the lever, centre to centre of the pivoting pins, and multiplying it by the chord factor given in the table below.

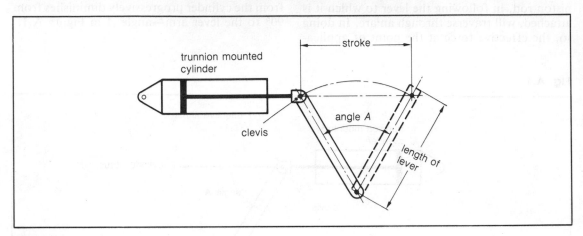

Fig. A.2

Chord factors

Angle A (degrees)	Chord factor	Angle A (degrees)	Chord factor	Angle A (degrees)	Chord factor	Angle A (degrees)	Chord factor
5	0.087	45	0.765	85	1.351	125	1.774
10	0.174	50	0.845	90	1.414	130	1.813
15	0.261	55	0.923	95	1.475	135	1.848
20	0.347	60	1.000	100	1.532	140	1.879
25	0.433	65	1.075	105	1.587	145	1.907
30	0.518	70	1.147	110	1.638	150	1.932
35	0.601	75	1.217	115	1.687	155	1.953
40	0.684	80	1.286	120	1.732	160	1.970

Example

A cylinder will be pushing a lever, 200 mm long, through an arc of 70°. The angle A will be divided equally by the perpendicular from the lever pivot point. What will be the length of cylinder stroke?

From the table the chord factor for an arc of 70° is 1.147. Therefore the stroke will be:

$$200 \times 1.147$$
$$= 229.4 \text{ mm}$$

8. Sizing a cylinder for a toggle action application

The force derived from a toggle action may be determined by a simple formula

$$F_t(\text{toggle force}) = \frac{F(\text{cylinder thrust}) \times A}{2B}$$

Looking at Figure A.3, it will be seen that A, although not the precise lever length, may be used as such for high leverage calculations with only a small error, since the lever is almost vertical. B is the distance between finishing stroke and vertical line of force.

Example

A cylinder with a thrust at 550 kPa of 1.73 kN is applied to a toggle action in which dimension A is 300 mm and dimension B is 10 mm.

What will the resulting force from the toggle action be?

$$F_t = \frac{1.73 \times 300}{2 \times 10}$$
$$= 25.95 \text{ kN}$$

Fig. A.3

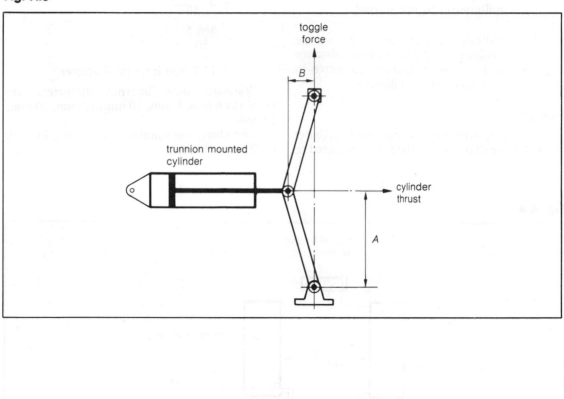

9. Selection of pipe size required between oil line cylinders and dash-pots

The following is a useful formula to use in obtaining a close approximation of the pipe size required between dash-pots and cylinder for any speed selected as the desirable maximum:

$$D_t = \frac{\sqrt{V}}{50}$$

where

D_t = diameter of minimum pipe size in millimetres

V = volume in cubic millimetres of displacement oil required to move the cylinder the required distance in millimetres in one second

or

V = cylinder piston area in square millimetres times required distance (i.e. distance required to move in one second) in millimetres.

Example

A hydropneumatically operated cylinder with bore diameter of 100 mm is required to move at a minimum speed of 100 mm per second. To find the pipe size connecting the cylinder to the dash-pots:

$$
\begin{aligned}
D_t &= \frac{\sqrt{V}}{50} \\[2mm]
&= \frac{\sqrt{\dfrac{\pi \times 100^2}{4} \times 100}}{50} \\[2mm]
&= \frac{\sqrt{\dfrac{22}{7} \times \dfrac{100 \times 100}{4} \times \dfrac{100}{1}}}{50} \\[2mm]
&= \frac{\sqrt{785\ 714}}{50} \\[2mm]
&= \frac{886.5}{50}
\end{aligned}
$$

= 17.7 mm internal diameter

Standard pipe internal diameters are normally 6 mm, 8 mm, 10 mm, 12 mm, 20 mm, 25 mm.

Therefore, the suitable pipe for the job will be 20 mm ID pipe.

Fig. A.4

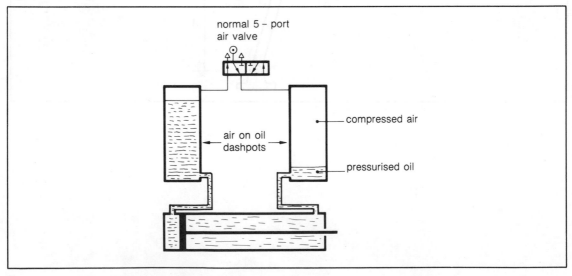

normal 5 – port air valve

compressed air

air on oil dashpots

pressurised oil

10. Selection of pipe size in relation to rate of flow, pressure and pressure drop

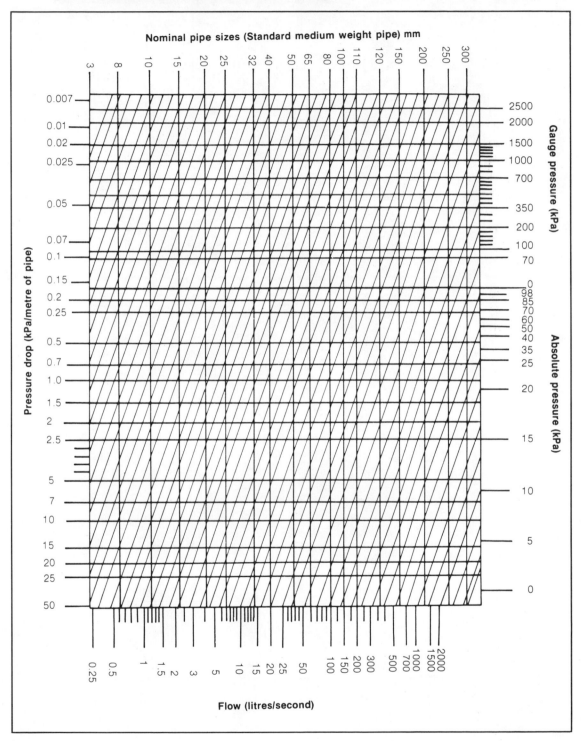

Fig. A.5

Index